Circularity

A Common Secret to Paradoxes,
Scientific Revolutions and Humor

Circularity

A Common Secret to Paradoxes, Scientific Revolutions and Humor

Ron Aharoni
Technion, Israel

NEW JERSEY • LONDON • SINGAPORE • BEIJING • SHANGHAI • HONG KONG • TAIPEI • CHENNAI • TOKYO

Published by

World Scientific Publishing Co. Pte. Ltd.
5 Toh Tuck Link, Singapore 596224
USA office: 27 Warren Street, Suite 401-402, Hackensack, NJ 07601
UK office: 57 Shelton Street, Covent Garden, London WC2H 9HE

Library of Congress Cataloging-in-Publication Data
Names: Aharoni, Ron.
Title: Circularity : a common secret to paradoxes, scientific revolutions and humor /
by Ron Aharoni (Technion, Israel).
Description: New Jersey : World Scientific, 2016. | Includes bibliographical references.
Identifiers: LCCN 2016000702 | ISBN 9789814723671 (hardcover : alk. paper) |
ISBN 9789814723688 (softcover : alk. paper)
Subjects: LCSH: Mathematics. | Philosophy. | Science--Humor. | Paradoxes. | Logic.|
Classification: LCC T49.5 .A338 2016 | DDC 601--dc23
LC record available at http://lccn.loc.gov/2016000702

British Library Cataloguing-in-Publication Data
A catalogue record for this book is available from the British Library.

Printed in Singapore

Cats and Tails

Before I start speaking, let me say a few words.

(An opening to a speech)

My little daughter returned one day from a visit to the dentist, and asked me: do you know how to make the anesthetic shot painless? You give an anesthetic shot before. Of course, to make this one painless, another shot is needed.

Amusing? Indeed. But strangely, this mode of thought also has serious sides. It is called "circularity", or "self-reference", and it is the protagonist of our story. The anesthetic shot is an example of a circularly defined task, namely one whose performance depends on its having been carried out before. Another example is the advice given to somebody who has fallen into a deep pit: go, fetch a ladder, and climb out. Such a situation is called, for good reason, a "vicious circle". The viciousness is experienced, for example, by a visitor to the United States who wishes to obtain a credit card. He has first to show a history of paying debts, of which the almost only instance is having paid the bills of a credit card.

A cat chasing its tail, someone who needs his glasses to find his lost glasses, a fresh job applicant rejected because of lack of work experience — all these are attempting to perform circular tasks. Most will fail. Lifting yourself from a pond, together with your horse, by pulling at your hair, is a feat that only Baron Munchausen knows how to perform. Archimedes formulated this idea in "Give me a lever and a place to stand and I will move the earth." He was not merely boasting of the wonderful pulleys he invented, but also wanted to explain that there is no leverage point from which you can lift yourself.

Circular tasks of another type are those that get in their own way. A photographer's command "Everybody, smile!" usually has the opposite effect. Polls published before elections can overturn the results and invalidate their own predictions. In Joseph Heller's *Catch 22* the pilots may evade bombing missions if they can prove insanity, but Item 22 in the army code says that anybody trying to evade bombing missions is sane. "Don't listen to this advice" is difficult advice to follow.

Circularity can be a pest. This happens when it outwits us, and passes circularly defined concepts as valid. For example, if you assume that the expression "The number hereby defined plus 1" points at a real number, you get an absurdity — a number equal to itself plus 1. "The father of Napoleon Bonaparte" points at a specific person; "The father of the person defined in this sentence" does not, and if you assume it does, you obtain the impossible situation of a person being his own father.

In these instances the circularity is transparent, and nobody will fall for it. But sometimes it succeeds in disguising well enough to fool us, and then absurdities arise. This is the dark side of circularity, its subversive and unruly face, to which the first part of the book is devoted. We shall meet in it a baby-snatching crocodile and agreements that cannot be fulfilled; proofs for the existence of God (or of the Loch Ness monster), and two famous philosophical paradoxes originating in circularity — the Determinism–Free Will conundrum and the Mind–Body problem.

Another, illuminated side, of circularity emerges when it is overt instead of acting from behind the scenes. It sometimes happens that nature itself poses circularly defined tasks, and in this case it is important to realize this. When a task is impossible, the impossibility had better be recognized. The second part of the book is devoted to this facet of circularity. We shall learn about three mathematical discoveries born from recognition of circular phenomena. One is a theorem of Georg Cantor from 1878, stating that there is no largest set in the world. For every set there is a bigger one. Then we shall meet the greatest ever victory of circularity — Kurt Gödel's Incompleteness Theorem, that (roughly) states that not everything that is true can be formally proved. And then we shall learn about a conclusion of Alan Turing's from 1936, that there is no omniscient machine. In modern day terminology — there is no computer program that can solve all problems. If such a program existed, it could pose a problem that it itself could not solve — just as an almighty God could create a rock too heavy for Him to lift.

The two sides of circularity are not separate. The rebellious side and the useful side are intertwined and influence each other. Paradoxes are sometimes inspiration for mathematical proofs (as was the case in the proof of Gödel's theorem) and mathematical proofs are sometimes parodied by paradoxes. The dark side is a negative image of the illuminated one, and negative images can teach something about the original. Moreover, they have their own charm.

For those who are ready to go deeper, I added a few chapters in *Part VII For the Experienced Hikers*. In those chapters some technical points will be elucidated, and some more complex mathematical insights will be presented.

Finally, no justice can be done to circularity without mentioning its charming side — its capacity to amuse. It is an unending source of jokes, and a sure recipe for humorous effect.

> I always thought I was indecisive. But now I am not so sure.
>
> The pig goes to God and complains: "Everything bad is ascribed to me: gluttony, filthiness, laziness. Why me?" God scratches His head, and says: "Indeed, piggishness."
>
> How does the dictionary define "circularity"? — see definition for "circularity".

In the last part of the book I will tell about circularity in humor, and try to solve the riddle of what makes circularity funny.

Acknowledgments

My gratitude to Orna Shai, who accompanied the book since its early stages, to Shira Zerbib-Gelaki for useful comments, and to Sarai Sheinwald for the illustrations. Thanks also to my son Ziv, for contributing this sentence.

Contents

The Dark Side — Paradoxes

Part I: Magic

I hear something falling, said the wind.
Oh it's nothing, it's only the wind, reassured the mother.

(Nathan Zach, "I hear something falling", *Different Poems*)

An Elusive Crook

Would you pay good money to be deceived? If your automatic answer is "no", consider again: isn't this precisely what you would be doing if you went to a magic show? The magician would fool you, conjuring the impossible before your very eyes, and you would enjoy it. In fact — you would be literally charmed.

This part of the book is about magic — rabbits that pop out of hats, women who continue to smile though sawn in half. But instead of a magician's stage, the setting is lecture halls and books. And these are not eyes that are tricked, but brains. The cons occur in the realm of the mental. They are called "paradoxes". A paradox is a deception that makes us experience the impossible. An absurd conclusion follows from seemingly impeccable assumptions.

If somebody told you that it is now raining in London and at the same time the sun is shining from one end of the city to the other, without a trace of a cloud, you would probably doubt his sanity. A paradox does precisely this. It poses a pair of arguments that seem both irrefutable, and yet they are contradictory. Of course, this must be the outcome of some cheating. In reality there are no contradictions, the world is consistent. It is some faulty assumption that makes us experience the impossible: a hidden mistake that leads to an overt absurdity.

In most paradoxes the deception is simple and is easily exposed. The listeners happily join in the laughter at their expense and go about their business. But some paradoxes have persisted for centuries, even millennia. And here something strange transpires: that all stubborn paradoxes are bred by one and the same trickster — circularity. In some mysterious way it convinces us that we can lift ourselves up by pulling at our hair. How does it perform this trick? Its secret is in being a master of disguise. And its craftiness is made easier by the fact that our brain is used to look outside, not inside, and therefore it has a hard time detecting it.

Here is for example a paradox named after its inventor, G. G. Berry, an Oxford librarian. Berry told the paradox to the mathematician-philosopher Bertrand Russell (1872–1970), knowing the latter's penchant for paradoxes. Russell published the paradox in 1906. Here it is:

> How many natural numbers can be described in English using fewer than a hundred letters? Surely many: "thirteen", "a million", "the number of inhabitants of China", "a million to the million to the million" — all these are expressions containing less than one hundred letters, and all define numbers. But still, there are only finitely many such numbers, because there are only finitely many expressions containing fewer than a hundred letters. Since numbers are infinite, there must be some that are not of this type. So, there is "the smallest natural number that is not definable by fewer than a hundred letters". But lo and behold — we have just defined this number, using fewer than a hundred letters!

A paradox? So it seems. But in fact this is only a disguise, not even a very sophisticated one, of a much simpler "paradox". Look at the following "definition":

The smallest number that is not defined by this sentence.

Which number is this phrase referring to? Is it 0, the smallest natural number? No, because then 0 is defined by the sentence, so the smallest number not defined by it is 1. Is it 1, then? No, because then the smallest number not defined by the phrase is 0. In fact, the phrase is obviously paradoxical: assuming it refers to a number, this number is by definition different from itself. Of course nobody would take this "definition" seriously. It is obviously self-referential. The circularity is too transparent.

But the truth of the matter is that Berry's paradox does the very same thing. It too defines a number as "different from itself". Using fewer than a hundred letters, it defines a number as "different from all numbers definable in fewer than a hundred letters" — in particular, different from itself (for it is one of those numbers that are definable in fewer than a hundred letters). How is it then that unlike "the smallest number different from the one hereby defined", Berry's paradox attained respectability? The secret is in diversion. More players are added to the game: the numbers two, four, the number of inhabitants of China — all those numbers that are definable in fewer than a hundred letters. All these are not really relevant. They are but a forest in which this single tree — "the number that is different from itself" — is hidden. And this thin shell of disguise suffices to make the definition look valid, and makes us wonder at the fickleness of logic.

A Nation Enamored with Paradoxes

A census conducted in 307 BC revealed that in Attica, the region surrounding Athens, there were 21,000 free men and about 400,000 slaves. The Greeks were not as cruel to their slaves as the Romans. They didn't make them fight wild beasts or each other in arenas, and a master couldn't take his slave's life with impunity. But the above numerical ratio meant that a free man could go through life without doing a day's work. Plato and his student Aristotle gave this state of affairs ideological legitimacy, which may shock present day enlightened people. The justification for human existence, so they claim, is thinking, because this is what sets humans apart from beasts; there are people capable of thinking, because they were born so, and there are people who are not, also because this is the way they were born; the latter are destined to do menial work, and are not on a much higher level of development than animals. The only justification for their existence is their usefulness in ensuring the continuity of the human race. This is true also for women, also workers in some sense, as they produce children. Slaves are more fortunate than menial workers, since they have the privilege of being in touch with their masters, and so enjoy the occasional brush with thinking beings.

Egocentric, chauvinistic and heartless? Perhaps. But there is truth in the proposition that slaves' labor played a crucial part in the miracle of the Greek civilization. Those of the Greeks that were free of work could use their time to think. And this they did, with great fervor. Another factor contributing to the miracle was the Greek admiration for abstractions. The world of thought was for them more important than mundane reality.

The big innovation the Greeks brought to science was the renouncement of practicality. Science was for them not a means but an end. Their predecessors, the Egyptians and the Babylonians, also studied mathematics, but with practical purposes in mind: measuring land in the Egyptian case and astronomy (important for their religion) in the Babylonian. The Greeks were the first to study mathematical concepts for their own sake. A fact like "there are infinitely many prime numbers" did not have any use at that time (nowadays prime numbers have direct practical value, for example in coding), but it ignited their imagination. Due to this approach, the achievements of the Greeks would nourish European civilization for the next two and a half millennia. For example, they bequeathed to the world the notions of axioms and proof, and the notion of constructions with ruler and compass. The Greeks were also the fathers of that least practical of all human endeavors: philosophy. Remoteness from applicability was for them proof of freedom from the shackles tying man to the material world.

And speaking of impractical games, what is less useful than a paradox? At least on the face of things, it cannot be of any practical use. It does not reflect on reality, because in reality there are no contradictions. A paradox is just a play with ideas. Somebody tangles a coil of threads, and then calls his fellow men to help him untie the knot. The Greeks loved to play such games. They came up with many paradoxes, some of which are discussed to this very day.

But the truth of the matter is that games are not devoid of value. Sometimes they have their moral. Paradoxes are useful in at least one respect: they force us to impose order on our thought. When a contradiction arises as if from nowhere, or an impossible conclusion is drawn from seemingly impeccable assumptions, we feel obliged to put the world back in order. An uncomplimentary mirror is put in front of our thinking, making us realize that we do not understand something elementary. We are left with no choice but to examine the hidden assumptions that lead to the error. Indeed, some of the Greek paradoxes were the gateway to deep insights.

Zeno and Tortoises

So many Greeks inhabited the foot of the Italian boot in the 6th century BC that the region gained the name Greater Greece. The best-known citizen of the provincial region was Pythagoras, a mathematician and philosopher, who established in the city of Crotone something unknown before or after: a mathematical cult. Around the year of his death another famous Greek philosopher was born there: Zeno of Elea (490–430 BC). He was preoccupied with questions of motion, and wrote a book containing forty paradoxes on motion. The book is lost, but some of the paradoxes survived to this day. The best known of those is that of Achilles and the Tortoise.

Achilles, one of the heroes of the Trojan War, is racing against a tortoise. Being fair, he gives the tortoise a head start, say one meter. Zeno's claim is that this will carry a heavy price: Achilles cannot win the race. Why? Because he first has to reach the point where the tortoise was at the beginning of the race, namely he has to run the meter's head start to get to where the tortoise was before. During that time, the tortoise moves a bit. Not much, admittedly, but he does make some progress. Next, Achilles has to get to this new location of the tortoise. While he does so, the tortoise progresses some more, a minute amount, indeed, but non-null. And so it continues: whenever Achilles gets to the point that the tortoise has just been, the tortoise goes a bit further. Conclusion? Achilles will never catch up with the tortoise. Of course, this is not what happens in reality, as can be easily realized by racing a tortoise.

This is one of those cases in which the deception is easy to expose. It is in the assumption that if we divide a time interval into infinitely many parts then this interval is infinite. In this case, the time lapsing until Achilles catches up with the tortoise is divided into infinitely many parts — until he meets the tortoise the first time, then the second time, and so on. But in fact, it is possible to divide a quantity into infinitely many parts and the quantity can still be finite.

To see this, let us look at another tortoise.

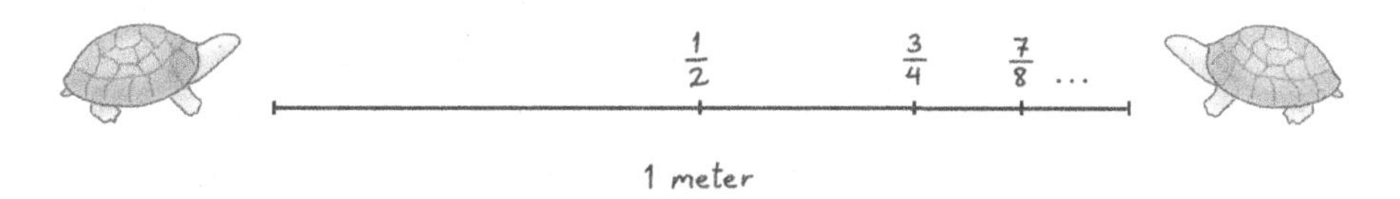

The tortoise in the picture wants to reach his beloved, who is at a distance of one meter from him. In the first minute he walks half a meter. This tires him, and in the next minute he walks only a quarter of a meter. Being then even more tired, he moves a one-eighth of a meter in the next minute. Of course, he will never reach the female tortoise. But does it mean that he walked an infinite distance? Of course not. The one meter distance is divided into infinitely many parts that are smaller and smaller. In fact, every high school student knows this secret. What does it mean that $\pi = 3.14159...$? It means that $\pi = 3 + 0.1 + 0.04 + 0.001 + 0.0005 + 0.00009 ...$. This is a sum of infinitely many numbers, but of course π is a finite number, in fact smaller than 4.

What happens here is that Zeno partitions the time interval that it takes Achilles to catch up with the tortoise into infinitely many parts. Let us make the calculation, and for this purpose assume that Achilles runs ten times as fast as the tortoise: he runs 10 meters a second, and the tortoise 1 meter a second (this must be a race tortoise). The first meter will take Achilles 0.1 seconds to run, in which period the tortoise would cover 0.1 meters. Achilles will run this distance in 0.01 seconds, in which period the tortoise will run 0.01 meters. Going on like this, we see that Achilles will catch up with the tortoise in

0.1 + 0.01 + 0.001 + ... = 0.111... seconds. This is an infinite decimal fraction, but its value is finite — in fact 1/9 (since its multiple by 9 is 0.999... , which is 1). It will take Achilles only one ninth of a second to catch up with the tortoise.

This paradox and its likes were instrumental in developing the mathematical thought about infinity, and about sizes that tend to zero. Eventually, these concepts ripened into the most useful mathematical tool, second in importance only to numbers, the infinitesimal calculus.

Are there Sand Heaps in the World?

Paradoxes had at the time of the ancient Greeks their own bible, a list of seven paradoxes collected by Eubulides, a philosopher who lived in the city of Miletus in the 4th century BC. They were the focus of heated debate at the time, and two of them still are. These are the Heap Paradox and the Liar Paradox.

In one of its modern formulations, the Heap Paradox is that you can pack infinitely many ties in a suitcase. How do you show this? Suppose that you managed to pack a certain number of ties. Can't you pack just one more? Of course you can. The tie is so small compared with the suitcase. So, you can pack more and more, *ad infinitum.*

This is of course a cheat. The answer to the question "Can't you?" is plainly — indeed you can't. When you can no longer stuff another tie is not clear, but at some point not even one more will go in.

The original formulation is more subtle, since it relates to thought and not to material things. It says that there are no sand heaps in the world. Here is the argument why. One grain of sand does not constitute a heap, that's clear. Now suppose that a certain number of grains do not form a heap. Adding one grain of sand of course will not turn it into a heap. Just one grain cannot make the difference between a non-heap and a heap.

But I can testify that in the building site next to my house there is a big heap of sand. There are many sand grains there, but only finitely many. There are heaps in the world.

I don't know what you think about this paradox, and of the fact that adults devote their time to it (some even being paid to do so).

Like all other famous paradoxes, it is born of circularity, although in this case the circularity appears indirectly. How can we know that? Here is one secret about circularity: how to identify it. There is a very simple means — breaking the circuit. If detachment makes the problem disappear, then it is a sign that its conception was in the sin of circularity. And if you ask how to detach, this is simple too: ascribing the problem to another person. In the case of sand heaps, think of another person, not you, who is looking at sand grains and heaps. You can then study the way the concept of the "heap" operates in his brain. You will probably find that from a certain number of grains he calls a collection of grains a "heap", and that this number depends also on location, illumination and the person's state of mind. All this is plain description (a very boring one, at that), and it does away with the paradox.

In what way is self observation different? In that when a person learns how the concept of the heap operates in his or her mind, he would remonstrate: "I don't care how the concept behaves. I want to know how it **should** behave, namely what is a heap **really**." This is the protest of the described against the description — a circularity phenomenon to which we shall return later.

Crocodiles and Lawyers

In the next story, taken from the book *Academica* by the Roman orator and philosopher Cicero, circularity appears directly. Its protagonist is the famous Greek sophist Protagoras. The sophists, who were active from the 5th century BC, were mentors for the art of argumentation. At that time being a lawyer did not require certification, and lawyers were chosen for their orating abilities. The sophists served therefore also as teachers of the law.

> A lawyer and his disciple signed the following agreement: if the student wins his first case, he is going to pay his teacher full tuition. If he loses then he is exempt. Having finished his studies, the student announces to his teacher that he is not going to pay him a penny, whatever the outcome of his first case. Of course, the teacher sues him. Now, this is the student's first lawsuit, meaning that if he wins it he has to pay the tuition. But "winning" means not having to pay! On the other hand, if he loses then he does not have to pay the tuition. But "losing" means that he does have to pay!

A paradox? So it seems. The student pays if and only if he does not pay, which is impossible. But a little contemplation shows that this is not a real contradiction. It is not that two contradictory things happen in reality, but only that the lawyer and his student made an agreement that cannot always be fulfilled. There is a situation in which it is self-defined. If the student's first lawsuit is between his teacher and him then the definition of the agreement is "you win if and only if you lose".

In the next story, from the time of Protagoras himself, there is another promise that is not realizable. The condition for its satisfaction is defined by its very fulfillment, in a negative sense. As if somebody is telling you "I promise not to fulfill this promise".

> A crocodile snatches a baby from its mother's lap. "Tell me something," says the cruel beast to the mother. "If you tell me a true fact, I will devour the baby. If you tell me a lie, I will drown it." The clever mother answers — "You will drown the baby." Now, if the crocodile devours the baby, the mother had lied, so he should have drowned it. If he drowns it, she told the truth, so in order to fulfill his promise he must devour the baby.
>
> For whoever believes in the integrity of crocodiles (why not, if you believe they can speak), the story ends well: the befuddled crocodile has no choice but to return the baby to its mother.

This, too, isn't a paradox. The crocodile's promise just cannot be fulfilled in certain instances. There are cases in which fulfilling it becomes a self-defined task. Not every promise can be realized.

Epimenides and the Mishtakes

Our next paradox does not involve an unrealizable promise, but a seemingly rock-solid assumption: that every statement about the world is either true or false. Sentences come into the world pre-labeled "true" or "false". But sometimes the label is so well concealed in the hem of the birthday suit that the suit has to be turned inside out to reveal the truth value of the sentence.

As we shall soon realize, this is not strictly so. A sentence has to earn its truth value by honest work, or more accurately — by examination. And sometimes the task of examining the truth value is circularly defined. In such a case, the sentence just does not have a "true-false" tag. But let us get to the story itself. Look at the following sentence:

This sentence contains two mishtakes.

The sentence contains a spelling error — this is one mistake, not two. That being so the claim that it has two mistakes is false. So there are indeed two mistakes. So, the counting of mistakes was correct, so there is only one mistake. And so on.

If you feel dizzy, like a cat chasing its tail, I achieved my goal — this is what circularity does. We entered an unending circle, also known as "infinite regression". We make one step then another, hoping to reach safe haven and never do. The result is a real paradox: there is one mistake if and only if there are two.

This is a variation on a simpler paradox, the Liar, which speaks about just one mistake:

L: This sentence is false.

If L is true, then by its content it is false. So, L cannot be true. It must be false.

On the other hand, if L is false, then it is true, so it cannot be false. It must be true.

We have shown that L is both false and true — a paradox. Later (following some medieval thinkers) we shall use the fact that L is both false and true.

An early version of the paradox is attributed to Epimenides the Cretan, who claimed that "All Cretans are liars". This does not lead to a real paradox: it is possible that Epimenides is lying, and somewhere in Crete there hides a truth speaker, not Epimenides himself. But the sentence L does seem to pose a true paradox.

Generations of thinkers have taken the paradox with utter solemnity. The engraving on the tomb of the Greek philosopher Philetas of Kos (340–285 BC) says that he died of lack of sleep trying to solve it. Is it really worth missing sleep over? Like every paradox, it is based on deception, and in this case clearly one that involves circularity, since L refers to itself. Indeed, there is a self-defined concept in it: L's truth value. The task of pinning a "true-false" tag to L is circular.

Think how you calculate the truth value of the sentence "My cat is black". To determine its truth value, you check the color of the cat. If the cat is black you write on the tag "true", and if not you write "false". Ascertaining "truth" means comparing the sentence with the world, specifically with the object of the sentence. Examine people who determine truth value of sentences, and you will find that this is what they do. Try to write a computer program that determines truth values, and you will realize that this is the way you have to write it. People, or computer programs, have algorithms in their minds, which decipher sentences, and compare them with fragments of reality. This, by the way, was Aristotle's view of "truth".

What is the object of L with which we have to compare it? L says something about its own truth value — that it is "false". So, the object is its truth value. As usual, in order to be able to compare the sentence with its object, you first have to know what this object is. In the case of L, you first have to know the truth value of L. But in order to know the truth value of L you have to calculate this very truth value.

So, the task of determining the truth value of L is circular. It is self-defined. The truth value is defined by itself. In fact, it is defined as "the converse of itself". Just like the number that is defined as itself plus 1. The truth value just isn't there. It is not well defined.

If it is so simple, then why is the paradox so persistent? Why do philosophers refuse to leave the carousel they are riding? The answer is that they decline to accept that the concept of "truth" is that simple. The "correspondence theory" of truth, that in order to know whether the sentence "The cat is black" is true you have to check the color of the cat, is the subject of a dispute, one of the best known in philosophy.

For example, one school of "truth" definers claims that the concept of "truth" is redundant. Saying "the sentence 'The cat is black' is true" is nothing but repeating the sentence "The cat is black". The two sentences convey precisely the same information. The word "true", says this theory of truth, doesn't add anything. And if this is so, then there is no procedure that determines the truth of the sentence.

The redundancy theory of truth is based on fear of circularity — the feeling that the definition: "The sentence 'A is true' is true if A is true" is self-defined. But of course, had the concept of "truth" been indeed redundant, it would not have been invented. "The sentence 'The cat is black' is true" speaks about a sentence, not about the cat, and thus it does contain information beyond the information on the color of the cat — if nothing else, that a sentence has been uttered. Sometimes you have to say things about sentences: they are there and they need to be recognized.

All this tail-chasing calls for some comic relief. Here is a joke version of the Liar Paradox.

I never make mistakes. I once thought I did, but I was wrong.

A Useful Paradox

Would you like to watch a logician at work? Here he is, Winnie-the-Pooh.

> "And if anyone knows anything about anything," said Bear to himself, "it's Owl who knows something about something," he said, "or my name's not Winnie-the-Pooh," he said. "Which it is," he added. "So there you are."
>
> (A. A. Milne, *Winnie-the-Pooh*)

"If (X) then I am William the Conqueror" you say when you mean that X is impossible. Pooh was not the first to employ it in order to "prove" what he had set out to prove. Medieval theologians used it in a more sophisticated way to prove the existence of God. And for this purpose they used the Liar Paradox.

The first to do so was the Parisian Jean Buridan (1300–1358). A very unconventional clergyman, he began as a womanizer and adventurer, which he later mixed with a not insignificant scientific career. For example, he was the first to speak about inertia — the fact that a body in motion will continue in a straight line unless some force is exerted upon it. He refused to join a religious order, but nevertheless became a famous scholastic (the scholastics were theologians who tried to produce logical arguments to confirm the tenets of Christianity). Nowadays he is probably best known for Buridan's Ass, the protagonist of a story he invented. This ass is both thirsty and hungry. To its right is a stack of hay, and on its left a pail of water, but since it is just as hungry as it is thirsty it doesn't know which to turn to first and dies of both hunger and thirst.

Buridan also investigated the Liar Paradox and, combining this interest with his scholasticism, he invented a proof of the existence of God. He considered the following sentence:

B: If there is no God then L is true.

L is the Liar's Sentence: **this sentence is false**, remember? — it is both true and false. Since it is true, B is true: "If X then Y" is true whenever Y is true. "If today is Tuesday then London is the capital of England" is true whether today is Tuesday or not. On the other hand, L is also false. So, B is just like "If there is no God then Pooh's name is not Pooh" — but this time with full justification (we have just shown that B is true). Since Pooh's name is Pooh, this proves that it is impossible that there is no God. So, God exists.

Of course, had Buridan considered the sentence "If there is God then L is true" he would have proved the non-existence of God. But when it comes to proving the existence of God, people willingly twist their logic.

Smullyan's Island

Raymond Smullyan (born 1919) is an American mathematician, pianist and composer of puzzles. He is also a circularity aficionado, which is evinced in both his mathematical interests and in his puzzles. His best-known book, *What is the Name of this Book?*, contains witticisms and mathematical tit-bits that are mostly circularity based.

One of Smullyan's favorite themes is an island he invented, which is inhabited by two types of people: honest people who cannot utter a lie, and obsessive liars, who can tell nothing but lies. Smullyan uses them for the purpose of all kinds of puzzles. Here is a famous one:

> You reach a road junction and you wish to know which road would take you to town, by means of a single question. You do not know whether the man you are asking is honest or a liar. What would you ask? One option would be to ask him, "If a person of the opposite type to yours was asked what the road to A is, what would he answer?" Take the answer and do the opposite. If he is a truth teller, then a person of the opposite type is a liar, and he would give the wrong answer, and this answer is what the truth teller would give; if he is a liar, then the person of the other type would give the right answer, but the liar will reverse it and again say the wrong answer. Another possibility is to ask: "If another person of your type was asked what the road to A is, what would he answer?" In this case, you should follow the road he directs you to (again, a bit of thought would reveal why).

In another story, Smullyan formulates a juicy version of Buridan's argument. One day, so his story goes, an islander is arrested on suspicion of murder. In his trial he claims the following: "The murderer is a liar". This piece of evidence by itself should suffice to acquit him. Why? Think: if he is honest then his testimony is valid and the murderer is indeed a liar, which means that it is not he (we are assuming in this case that he is not a liar). If he is a liar, then, since his testimony is false, the murderer is not a liar, so again, it cannot be he.

The problem is that the man was found beside the corpse with a bloody knife in his hand and a wide smile on his face. He is obviously the murderer, which means that he managed to prove an obvious fallacy. It seems that using his method, he can prove anything. And indeed he can. See what he is claiming when stating that the murderer is a liar: "If I am the murderer, then I am a liar", which means "if I am the murderer then this is a lie". In other words — "If I am the murderer then L is true". And as we saw in the last chapter, this proves that "I am not the murderer".

Smullyan relates that when he was ten his big brother entered his room on the morning of April Fool's Day, and told him: "Today I am going to trick you like you have never been tricked before". Little Raymond waited, and waited, and waited, and nothing happened. To this very day, he writes, he is not sure whether his brother tricked him or not. If he did not, then he did the opposite of what he promised, which means he did trick him. If his brother did trick him, then he fulfilled his promise, which in turn means that he did not trick him.

Proving the Existence of the Loch Ness Monster

Circularity is a strong weapon in the hands of swindlers. In their defense it must be said that usually they trick not only others, but also themselves. The author of one such swindle was Anselm of Canterbury (1033–1109), who was not born in Canterbury but in the Italian Alps. As a young man he joined the Benedictine Order and as part of his assignments he travelled to France and later to England, where he rose to the rank of the Archbishop of Canterbury. He was one of the first scholastics, and one of the most influential among them. Like all scholastics, he was ready to juggle with thought any which way to prove the existence of God. The one I am going to relate now was his most successful "proof".

> God possesses perfect attributes. Non-existence is an imperfection. Hence God, as we perceive Him, must exist. If He does not, then it is not God that does not exist but some other entity. The entity we defined as God does have existence among its attributes.

Of course, this argument can serve to prove the existence of anything we wish. For example, define the following creature: "The existing monster of Loch Ness". If you find out that it does not exist, then the thing that does not exist is not the monster we are speaking of. That one does exist.

The philosopher Immanuel Kant (1724–1804) named this argument the ontological (i.e. "of existence") proof. He was also the first to expose its fallacy. Existence, so he understood, is not an attribute of the object. It is a property of the concept. In the sentence "The cat is black" the word "black" relates to the cat; in the sentence "The cat exists" the word "exists" does not relate to the cat but to its concept. It says that there is something in the world that corresponds to the concept "black cat". If you want a proof that Kant is right, consider the sentence: "The cat does not exist." Obviously, "exist" here does not relate to the cat, because the cat isn't there. We are saying that there is nothing in reality corresponding to the concept. When we say that unicorns do not exist, clearly the "do not exist" doesn't relate to unicorns. It refers to a concept in our mind.

So, Kant is saying that the concept of "An existing God" is a result of confusion. The concept concerns its **own** relationship to reality, meaning that it is circularly constructed. A proper concept defines the properties of its object not of itself.

In fact, the structure of the concept in the ontological proof is very similar to that of the Liar Paradox. To see this, look at the following conjecture, call it "C":

C: There is a monster in Loch Ness, and this conjecture is true.

This conjecture must be true. For, if it is false, then it is another conjecture that is false: ours is defined as true! But the veracity of C means, in particular, that a monster exists in Loch Ness. So, we proved the existence of the monster!

And who knows, perhaps there is a monster there indeed. Just try telling a Scotsman that there is none.

Part II: Free Will

We must believe in free will — we have no other choice.

(Isaac Bashevis Singer)

Newcomb Overturns the Rules of Nature

Suppose that you know an infinitely wise person, who can predict events with utmost precision. He uses a super computer and very precise instruments that measure the human body and brain. Until now he hasn't failed even once in his predictions.

One day the super-predictor honors you yourself and predicts a choice you are about to make. He puts you before a well with a one hundred dollar bill in your hand, and offers you a choice: to throw the bill into the well or hold on to it. The sage tells you that he has already predicted your choice but is not about to divulge it to you so that you won't prove him wrong by intentionally choosing otherwise. He does tell you that if his prediction was that you throw the bill into the well he would have deposited $1000 in your bank account half an hour since. If his prediction was that you'd keep it, he would have done nothing.

The paradox lies now in the existence of two contradictory arguments. One is that whatever you do now will not alter what had been done a half hour ago. There is no point in throwing the bill since it cannot affect the past event of having money deposited in your account (or not, as the case may be). The other argument is that the predictor is always right. If you throw the bill, assuming that had been his prediction, he had deposited $1000 in your account — you gain $900!

This paradox was formulated in 1960 by the Californian physicist William Newcomb, who arrived at it while investigating a problem in mathematical game theory. It took his friends nine years to convince him to publish the paradox, but when he did it started what the aficionado Isaac Levy called *Newcombmania.* Many have contracted the disease and hundreds of articles have been written on the paradox. Indeed, there is no denying its attraction.

Martin Gardner, a famous math popularizer and author of mathematical riddles, claimed that all solutions of the paradox offered just vehemently repeat one of the two arguments. This, of course, is useless. There is no point in supporting one of the arguments — the problem is that they **both** look correct. What is needed is to understand why one of them is not.

The point of bifurcation, where the ways of the two arguments diverge, is clear. One argument, the one against throwing the bill, states the firmest of all human beliefs, that the past cannot be changed. No present decision can affect past events. The other argument is no less than a refutation of this very belief. Not merely a refutation, but a very concrete one. It points at an explicit way for doing precisely what the first argument claims to be impossible: determining the past, in the sense of doing something in the present that determines the fate of past events. Using his sage, Newcomb establishes a causal connection between the throwing of the bill, which is done now, and the amount in your account, determined half an hour ago. This causal link can be used to decide the fate of past events.

Isn't this astounding? Newcomb has apparently discovered the philosopher's stone. Who wouldn't give all he has for the ability to change the past? Deciding the fate of the past is one of mankind's deepest desires. Much more so than turning lead into gold. Wizards — go to Newcomb and learn his ways. You have a lot to learn about conjuring the impossible!

Who is right, Newcomb or the intuition that the past cannot be altered? Is Newcomb's trick valid? If so, then the belief that the past cannot be determined by present decisions is a mistake, the result of a blind spot in humanity's field of vision. We have just not been alert enough to discover the secret. We all lived in the dark, until Newcomb came and showed us the light. On the other hand, if it is indeed impossible to change the past, as we have all believed until now, then there is some fallacy in Newcomb's argument. He is cheating in some subtle way.

In order not to keep the reader in suspense, let me tell you right now: Newcomb is cheating. He is guilty of a circular argument. But the analysis leading to this conclusion is not straightforward. In order to present it, I need to go on to another, much better known puzzle.

Written in the Stars

> A slave misbehaved and his master was about to give him a thrashing. "It is not my fault," said the slave, who knew his master's philosophical penchant. "It was written in the stars that I should sin." "Yes," the master returned the obvious answer, "and it was also written in the stars that I will thrash you."

This story, from the Roman times, illustrates a famous riddle — the problem of free will. In fact, it is not merely a riddle but a paradox. Two arguments are presented, each apparently trivially obvious, and yet they contradict each other. Like many other famous philosophical problems, Plato was the first to formulate this one. The world is governed by laws, he said, so this must be true for human decisions as well. They too obey causal laws. Whether by psychological or by physiological laws governing the chemical reactions in the brain, human decisions are predetermined. Given the state of the world at the present moment and in particular the state of the brain of the decider, it is possible to know, at least in principle, what the decision will be. But if that is so, where does freedom of choice come in? Somebody else is making our decisions for us — the laws of nature. What is the point of attempting to resist what is written in the stars?

Many who dealt with this enigma related it to circularity. In Somerset Maugham's *Appointment in Samarra* a slave goes to the market of Baghdad, where he meets a woman whom he thinks had made a threatening gesture at him. Being sure that the woman is no other than Death, he asks his master for permission to run away to Samarra. Having granted the permission, the master goes to the market, meets the woman, and asks her why she frightened his slave. "I was not trying to frighten him," she answers, "I was just surprised to see him here. I have an appointment with him in Samarra."

The best-known story involving circularity on the uselessness of trying to evade fate (represented by the *moirai* in Greek mythology) is that of Oedipus. When he was born the Delphi oracle prophesied that he would kill his father and marry his mother. Terrified, his parents gave him to a shepherd, to get rid of him. This initiated a chain of events culminating in the fulfillment of the very prophecy.

This conundrum is compounded by the problem of responsibility, exemplified in the story of the thrashed slave. If man's choice is predetermined, then he cannot be held responsible for his actions, neither good — deserving prizes, nor bad — eliciting punishment. After all, it is not he who generated the causal chain leading to his actions, why praise or castigate **him**?

Another formulation of the problem comes from religion. If God has foreseen it all, then where is the freedom of choice? And if He made me sin, or at least didn't stop me from sinning, how can He hold **me** culpable? The Jewish *Talmud* (book of religious laws) engages with this problem, and a devious solution is offered by one of its interpreters: "All is in the hands of heaven except the fear of heaven" (*Talmud*, Berachot 33b). In Christianity the problem was discussed by the Egyptian-Greek philosopher and theologian Saint Augustine (354–430 AD).

The idea that everything is predetermined is called "determinism". "Predetermined" in two senses: one, that things can be predicted, and the other determination by laws. Of course, the two are connected because prediction relies on the laws. Democritus (460–370 BC), the founder of the atomic theory of matter, is considered the first propounder of the idea of determinism as well. In the 18th and 19th centuries, following the victories of the Newtonian theory, the deterministic view reached its pinnacle. Physicists viewed the world then as composed of molecules rolling about like billiard balls, and the motion of billiard balls can be predicted using Newton's laws. The mathematician-astronomer Pierre-Simon Laplace (1749–1827) inferred that from the current state of affairs one can deduce the state of the universe at any point in time, past or future.

Since then science has sobered up a bit. Quantum theory poses limitations on the precision of measurements and predictions. Thinking quantum, you can only predict microscopic events with good probability. Indeed, there are people (notably the physicist Roger Penrose) who claim that free will resides in the small quantum leaps in our brain. But this claim suffers from more than one weakness. For, don't we incessantly predict the actions of our fellow men? When I get on a bus I predict, usually accurately, the route the driver is going to take, and that he will not start singing aloud. Another weakness is that the decision-making processes in our brains involve non-negligible amounts of energy. Every decision maker knows how much energy they invest in deliberation. And the strongest point against this view is that people who have never even heard about quantum theory feel the contradiction between predetermination and freedom of choice. And the source of this intuition should be explained.

As in other paradoxes, the contradiction between free will and determinism doesn't give respite to people to whom such problems do not give respite. And as in other paradoxes, it is most likely born of some error. But, as we shall find, the deception here is more refined than those we have met with so far. Not in vain has the problem gained such a central place in philosophy.

Nothing to It — Or Perhaps There is Something?

Among serious philosophers I know it is agreed that when they read a new article not only do they find there annoying inaccuracies, but the entire idea looks to them wrong from beginning to end.

(Stanley Cavell, *Do We Have To?* p. 10)

The feeling of contradiction between free will and determinism is so strong, that it is hard to deny. It seems self-evident that predetermination of one's decision is incompatible with the feeling of freedom of choice. And yet, the reason for this incompatibility is more a matter of intuition than an ordered chain of arguments. For this reason, many have claimed that the sense of contrast is merely an illusion, the result of some misconception. Three famous philosophers made this claim, each in his turn: Thomas Hobbes (1588–1679), David Hume (1711–1776) and John Stuart Mill (1806–1873). All three ascribed the problem to an inane error, confusion between two meanings of the word "predetermination" — external coercion as opposed to being governed by rules. If somebody is forced to turn a millstone, he is indeed not free. But in the context of Determinism–Free Will, the determinism is not of this type. The actions considered there are still dictated by man's own will. The fact that God or Laplace (who as mentioned in the previous chapter claimed that all is foreseen) can predict a person's choice does not mean that the actions are not the result of that person's free choice, only that the choice is governed by laws. "All is foreseen and freedom of choice is granted", as the Jewish *Mishna* says (*Pirkey Avot,* chapter 3, *Mishna* 19[a]).

The claim of Hobbes, Hume and Mill is reminiscent of a famous claim made by Ludwig Wittgenstein that all philosophical problems, large and small, are based on erroneous conceptions. Moritz Schlick, Wittgenstein's disciple and heir in Vienna, asserted the same about the responsibility component of the problem. Lamenting the paper wasted, the ink spilled and the intelligence that could be used to solve other problems (not forgetting to add "assuming it would suffice for them"), he said that the concept of responsibility is simply a positive and negative incentive to action, encouraging some and discouraging others. It has nothing to do with whether or not the action was predetermined. In fact, it is precisely determinism that gives the concept of responsibility its justification. But for the existence of motives, there would be no point in prizes and penalties.

All this is fine, but nobody is convinced. The "all is vanity" claim has left the philosophical world indifferent. And yet it moves, feel most thinkers, and the problem refuses to be buried. Moreover, the objection seems justified. The sense that free will and determinism are contradictory is so strong and persistent, that to attribute it to some inane confusion does not seem plausible. Intuition must be pointing to something that needs explicit mention. Indeed, at the heart of the problem is an argument that cannot be lightly dismissed.

The Idle Argument

Not all components of the free will issue are dim. There is one argument that is transparent and direct, and takes the bull by the horns. It cannot be ignored nor ascribed to a misconception. It is a real paradox, as sharp as any we have met. It is called "The Idle Argument". Here is one way to formulate it:

(a) It is impossible to change the past. Present actions cannot cause past events.
(b) It is possible to change the future by present actions.
(c) Determinism means equivalence between past and future events.

"Equivalence" between two events A and B means that A happens if and only if B happens. For example the weather forecast "it will snow tomorrow" may be equivalent (if it's any good) to tomorrow's snow.

Put together, these three statements yield an absurdity: two events are equivalent, and yet one can be determined while the other cannot. Clearly, if events A and B are equivalent, then an action that determines the fate of A also determines the fate of B. If the action cannot affect B then it cannot affect A.

This argument goes back to Aristotle. Three hundred years later, Cicero, whom we met in the chapter Crocodiles and Lawyers, asked what the point of seeing a doctor was, if it is already written in the stars (or predicted by some super predictor) whether or not you will recover. You cannot change the prediction so you cannot influence the event of your recovery. (As an aside, I suppose that doctors then were indeed a bit of a waste of time.) Such stories are the origin of the name "Idle Argument": if your fate is known in advance, there is no point in struggling. For example, if my grade for tomorrow's exam is written in the stars, why should I sweat for it? The grade will be the same whether I do or not. Should you fall into raging waters, there would be no point in trying to swim. Some sage would have already foretold whether you will sink or be saved, and this you cannot change. Your fate is known in advance, so why bother to fight or why worry?

All this sounds convincing, but clearly no drowning person would be persuaded to stop swimming upon hearing this argument. And if he does, the sage would have predicted this decision, and written "drown" on his note. If you sense circularity, you are right. It will soon appear.

This argument gives explicit expression to the otherwise vague sense that free will and predetermination are contradictory concepts. It is an honest paradox, and no wonder: it is a reformulation of Newcomb's paradox, no less. This should of course be turned around: Newcomb's paradox is nothing but the idle argument revisited.

The starting points of Newcomb's paradox and the idle argument are the same: equivalence between past and future events. Event A in the past (the deposition of money into the bank account) is equivalent to event B in the future (the throwing of the bill). Now it would seem that the two arguments part way.

> The idle person says: since I cannot change A, I cannot change B.
>
> Newcomb says: Since I can determine the fate of B, I can also determine the fate of A.

Divergent ways? Not at all. A bit of thought reveals that the two arguments are identical, two different ways of saying the same thing. "If today is Wednesday then tomorrow will be Thursday" says precisely the same as "If tomorrow will not be Thursday then today is not Wednesday". "If X then Y" is the same as "If not Y then not X".

Moreover, in both arguments the problem does not lie in the mere acknowledgement of the equivalence of past and future events, but in an operative conclusion: "Therefore, it is worthwhile doing this, that or the other." Newcomb says, "Therefore it is worthwhile throwing the bill." The idle argument says, "Therefore it is not worthwhile preparing for the exam." In both cases it is the operative conclusion that makes the paradox. Without it, you are just aware of some causal chain of events. There is no altering the past.

This means that if we understand the flaw underlying Newcomb's paradox we shall also solve the paradox of the idle argument. And since the idle argument is at the core of the free will problem, this latter can also be settled.

The next question is whether circularity is involved here, as would be expected in a paradox. We already know how to test this assumption — by separation. If the problem disappears when viewed from the outside, then its origin must be in circularity. And indeed, separation makes both problems disappear. The entire point in Newcomb's paradox is in the decision "Therefore it is worthwhile to throw the bill." This cannot be said by a detached onlooker, because an onlooker is not in a position to decide. You can no more decide for somebody else than you can eat or sleep for them.

Looking from the outside, relinquishing the bill does not **cause** the deposition of the money in the bank. The two are merely causally linked. An onlooker may wish that the chooser will throw the bill but he cannot transform this wish into a decision. He can use the causal link to serve his own decisions, for example for betting. If he believes with 99% probability that the sage was right, he can put 99 dollars against one that there will be money in the account if and only if the chooser will throw the bill. Stranger still: the bill thrower himself can regard himself in a detached manner and make the same bet. This still doesn't constitute "affecting the past".

The disappearance of the paradox upon separation is a clear sign that it harbors circularity. What this circularity is — is not yet clear.

Why Can We Not Change the Past?

It's a poor sort of memory that only works backwards.

(The White Queen in *Through the Looking Glass*, Lewis Carroll)

In order to resolve the question on which side the deception is — Newcomb's or the conviction that we cannot affect the past, we need to know where this conviction stems from. One clue is that it is not an isolated belief. It is one of a pair of Siamese twins. The other twin is another asymmetry of time, which we also accept as self-evident (except for Lewis Carroll, see above): that we cannot remember the future. I know with certainty what I ate for breakfast this morning; as for tomorrow's breakfast — I cannot remember what I will have eaten then.

On the face of it, affecting the past and remembering the future are totally different things. We usually don't even connect them. But this is an optical illusion. In fact, they are very similar: they are mirror images of one another. Both are links of our brains with events. Decision is a link with the future and memory with the past. And both have the same pattern: a brain process links to some external event through the mediation of a body process. In the case of decision, the brain process is the deliberation, and the body process is the action it entails. In the case of memory the body process is the sensations that past events have inscribed upon the brain.

Here is a figure that demonstrates the symmetry:

future event

physical process (action)

brain process (deliberation)

present

direction of time

brain process (memory)

physical process (sensations)

past event

Why can't we see the similarity between remembering and deciding? Why do we view them as so different? The reason is our asymmetric view of time. Our relationship to the future is very different from that which we entertain for the past. And the reason for that is simple: that we are animals.

Animals and plants are future oriented beings. Their eyes are stuck in the front of their heads (if they have heads), in the direction they are heading towards, not that from which they are coming. Even a historian, when preparing an omelet, thinks of where the egg is heading and not where it came from. Blaise Pascal, a bright mathematician in his youth and a philosopher later on in life, wrote in his book *Pensées*:

> *We almost never think of the past. And when we do, it is only to shed light on our relationship with the future.*
>
> (Blaise Pascal, *Pensées,* part 80)

The reason for this is that life as we know it is the result of evolution, which as a matter of tautology produces beings that act so as to preserve their species into the future. Animals and plants are programmed to affect their future ambience, aiming to make it favorable for survival and for reproduction.

The inanimate world has a different asymmetry of time, known as the Second Law of Thermodynamics (the first is the conservation of energy). It is that disorder (measured by a parameter called *entropy*) increases with time. The milk poured into a coffee glass starts in an orderly state — being concentrated in one place, and ends up in a less orderly state — mixed with the coffee. A house left unattended slowly disintegrates, and wear and tear processes will eventually bring it to dust. Energy invested from the outside is required to increase its order. Nobody knows for sure why the world is going from order to disorder, but this is so.

These two asymmetries are related. Evolution wouldn't exist but for the increase of entropy. It is a rebellion against the Second Law, being based on a marvelous trick that defies the increase of disorder: reproduction. Order is indeed rare, but if it is of a type that reproduces itself, it will become less rare. If materials from which a rabbit is formed are put and stirred in a cauldron, the chances that a rabbit will emerge are slim. But if by a very rare coincidence a male rabbit does appear, and in the adjacent cauldron a female rabbit is formed, they will both jump out of the cauldrons and start reproducing, thereby making the order typifying them less rare. Of course, this is not the way evolution happened. It did not rely on such slim chances, and it took more time.

This explains why causal links of an animal's actions with the future are so important to it. And why this animal, even if endowed with impressive verbal skills, will find it hard to see the similarity between its relationships with the past and with the future. It views the two as so distinct, that it cannot even discern the symmetry between them.

So, the answer to "why can't we change the past?" is simple: it is not a law of nature nor an axiom but a matter of definition. "Determining the fate of events" is what we call the relationship of our brains with the future, and "remembering" is what the past engraves on the slate of the brain. We cannot see it, because these definitions are engraved deep in our minds. In fact, even in our bodies.

And Yet It Moves?

Have we settled the problem? Not quite. For, it seems that somebody miraculously succeeded in designing a situation in which the past **is** affected: Newcomb. He describes a way of affecting the past via causal connections. Here is an illustration of this situation. The arrows in the diagram denote causal links, and as before, the arrow of time points up.

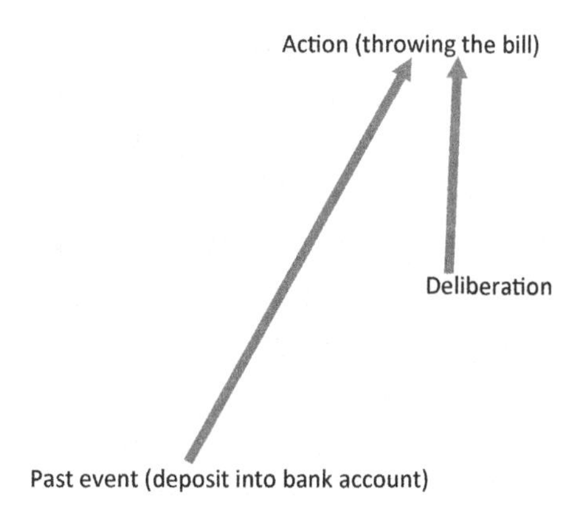

An important feature of this diagram is that the causal link between the past event and the action circumvents the process of deliberation. It arrives from aside. The point we shall reach in a moment is that this is quite impossible, but precisely for this point I drew the diagram. Now I will ask you to pause for a moment, and conjure in your minds such a situation, of a causal link between a past event and a choice, flanking the deliberation on this choice. Can you come up with such a link?

This is a useful exercise, and if you try you will probably find that it is not easy. In fact, the main benefit of such an attempt is the realization how unnatural such a situation is. Here is a suggestion: if the sage deposited money into your account, he whispers to a giant he finds around (where there are sages there are also giants) to take your hand and shake it so that you drop the bill. If he did not deposit the money, he tells the giant to prevent you from dropping the bill. The causal link now does not go through the bill-holder's deliberation but through some fierce giant. And since here, too, you will be richer if you drop the bill it is advantageous to do so!

"Advantageous"? Not quite. "It is best to throw the bill" is here an absurd conclusion. You are not deciding and you are not choosing. And this absurdity points at the problem in such a situation. When there is a causal link from aside, not going through your deliberation, you feel that you are not free to choose. The action is not given to your decision. Links of an action with the past that circumvent the process of deliberation are grasped as coercion.

This is precisely the observation of Hobbes, Hume and Mill, on the difference between coercion and predetermination. Coercion is a will-circumventing causal link, while predetermination involves causal links that go through the decision. Hobbes, Hume and Mill claim that the Determinism–Free Will problematics arises from an assumption of causal links arriving from aside, while in fact predetermination involves causal links that go through the decision.

Why do we view causal links from aside as coercion? This also has a deep reason. When deliberating an action, it is important for us to separate those elements that are given to choice from those that are not. For example, if I am contemplating a visit to the beach, I have control of the direction in which I carry my body. But other components, like gravity, speed limit on the highway and the weather, are not given to my decision. All these factors arrive from besides the deliberation, and pose constraints in my way. In order to decide, I need to know what I do control, namely what part of the action is given to choice.

Cat and Tail: Choosing the Choice

We now understand: the link between the future action and the past event cannot circumvent the deliberation. It must go through it or else the decider feels coerced. This is the reason that Newcomb uses a predicting sage and not giants. The predictor does not predict the action but the process leading to it. He is familiar with the motives and the way the decision maker's brain works. In short, he is predicting the deliberation. The causal link must look like this:

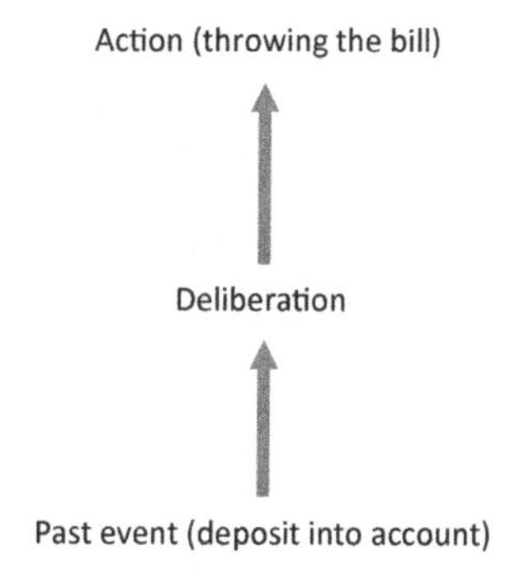

This is the real diagram. The prediction is of the decision process and only indirectly of the action.

From here the way is short to identifying the circularity in Newcomb's argument. Since the sage predicts the decision process, the argument for throwing the bill really says: "I should throw the bill, so that my decision process is such that the sage, upon understanding it, will have deposited money into my account." Which is, in fact: "I should throw so that my deliberation process will be so and so."

This means that the person using the argument for throwing the bill is trying to choose his choice process. He is deliberating the deliberation. In other words, he is trying to choose his motives. And this is a circular task. You can act upon your motives, not decide what they should be. You cannot put up for decision the decision process itself. Trying to do so is like trying to lift yourself by the tuft of your hair. And as we know, believing that this is possible readily leads to paradoxes.

Subtle as this argument is, it is the ultimate reason for the belief that the past cannot be changed. And it anchors this belief in firm ground. Trying to change the past inevitably leads to a circular task. The belief that the past cannot be affected is fully justified — there is no point in trying to alter yesterday's events.

The Matrix

In *The Matrix*, a successful science fiction film from 1999, humans are programmed by machines that are cleverer than they. This is less fictitious than it may seem: can anybody claim that he knows the deepest motives that make him tick? We are all programmed by our history and by our nature. In the movie, by the way, not as in life, the humans manage to break free, with the help of a benevolent machine.

This idea makes it possible to establish another connection between choices and past events: instead of through prediction, through programming. Suppose that a programmed person finds himself one day in front of a well with a hundred-dollar bill in his hand, and the machine that programmed him tells him: "Do you know, you are only a program. If I programmed you so that you will throw the bill, I will have deposited a thousand dollars into your account. If I programmed you not to throw it, then I would have done nothing." The dilemma is now the same as in the original paradox. Should the person decide to throw, so that he was programmed to throw, so that he would have a thousand dollars in his account? The programming happened in the past — how can he change that?

The solution is the same: a program can decide on many things, but not on the way it was programmed. Being in the place of the programmed decider, I would hope that I am that type of program that throws bills. I would then be richer. But I cannot "decide" to be of that type.

"To decide what your motives should be" — this brings to mind a man whose motives turned out to be more important than his action. Herostratus was a young man who decided to become famous for a big crime. On July 20th 356 BC, a day remembered also for being the birth date of Alexander the Great, he burned down one of the seven wonders of the ancient world, the temple of Artemis in his town Ephesus (now in Turkey). In his trial he boasted his deed and claimed that this way he would be immortalized. To foil his plan, the judges not only condemned him to death but also forbade, under pain of death, to mention his name. Ironically, he did become immortal, not for his deed, but for his motive. His name is now identified with crimes perpetuated in order to gain fame.

If, somewhere in afterlife, Herostratus knows that this is so, he is probably content. Could he then decide: "Let me burn the temple so that my motive would have been such and such, which will make me famous?" And wouldn't that be deciding on your motive? Not quite. If this were so, the real motive would not be the wish to become famous for the arson, but the wish to become famous for that motive. This would be a different motive, about which he does not decide. He is still not deciding on his real motive. A bit dizzy? So you should be. Infinite regression (deciding on the motive for the motive ...) is confusing.

All is Foreseen (From the Outside) and Permission is Granted (From the Inside)

Since the Idle Argument is virtually identical to Newcomb's paradox, the solution is valid for it too. The protagonist of this story is also deliberating, using a causal chain that includes the deliberation itself. His conclusion, "I'd better stay idle", uses a link in the chain that is this argument itself. Like the thrower of the bill, he is trying to reach a decision on his decision process.

The conclusion is that Hobbes, Hume and Mill were right. The conflict between determinism and free will is indeed imaginary, but due to a deeper error than the three had identified. The contradiction between determinism and free will does not exist, because the two do not apply to the same person. Determinism is observed from the outside, while free will is viewed from within. From the outside, everything is foreseen. From the inside, free will is granted. And since the two relate to different beings, they are not mutually exclusive or contradictory.

"The unexamined life is not worth living," said Socrates (470–399 BC). Probably a bit of an exaggeration: there are people who do not analyze their motives and yet lead a worthy life. There is a time to think and there is a time to act. When you cross a busy road, don't look at your motives but at the cars. In the last few chapters we learnt the limitations of self-examination. When a person is in the midst of a deliberation process, he cannot look backwards. The deliberation is a wave carrying him and letting him look only forward. Everybody knows: deciding which car to buy demands considering cars, not one's motives. The inability to examine your motives in full, while you are deliberating, is a matter of principle. The examination is bound to interfere with the motives, and thus it becomes a circular task. You cannot burn the candle at both ends — you are either an observer or a decision maker.

What we learnt in the last chapters is that there is an even deeper phenomenon of circularity here: not only knowledge of the precise nature of causal links of the decision with the past, but even the mere knowledge of their existence, cannot be used for the decision. This is the deep reason that we cannot affect the past.

Part III: The Mind–Body Problem

Only be sure that thou eat not the blood: for the blood is the soul;
and thou mayest not eat the soul with the flesh.

(*Deuteronomy*, 12:23)

A Meeting of Non-Meeters

A man loves a woman. Is there anything more abstract and non-tangible than love? It has no mass, temperature, color or energy. Nobody has ever seen "love". And yet the love of the man moves worlds. It causes cities to go to war, ships to sail, a huge wooden horse to invade a walled city. How can a non-physical event affect physical events so massively, or indeed affect them at all? How can a process not involving exchange of energy influence physical processes involving such massive exchanges?

The same question can be asked from the opposite direction. A physical event, like a needle's prick, can cause a mental one — pain. External events cause upheavals in our mind, not to mention the perish of souls. How can this happen? How can a physical entity be in touch with a non-physical one? How can two players that don't share the same playground play together?

This riddle is called the "Mind–Body problem". Like the Determinism–Free Will problem, it has the form of a paradox. An impossible connection is materialized before our very eyes. Many philosophers view it as the central paradox of the entire philosophy. The Austrian–New Zealand–English philosopher Karl Popper (1902–1994) wrote, "This is an axis around which all Western philosophy revolves." There is truth in it. This paradox summarizes the problematics caused by circularity in the most succinct way. As we shall see, it epitomizes what happens when a man looks at himself without separating between observer and observed.

Like all central philosophical problems, the seeds of this one date back to the Greeks, mainly sown in Plato's writings. It was first formulated explicitly by René Descartes, in his book *Meditations*, from 1641. Descartes also had an original solution: indeed, he claimed, mind and body are distinct entities, but they connect at a point in the brain called the hippocampus. Another courageous solution was proposed by Gottfried Leibniz (1646–1716), who besides being a great mathematician and an influential philosopher was also an energetic theologian. Mind and body are indeed separate, he said, but they were both set by the same authority. God set the two worlds, physical and mental, so that they are synchronized. It is just as if two clocks show the same time despite being far apart, because the same watchmaker set them. In philosophy departments these solutions are taught without a hint of a smile, but I do not expect that the reader will take them seriously. I am just trying to show how desperate philosophers are in tackling the problem.

You may not be surprised to hear that, as in the case of the Determinism–Free Will problem, there are philosophers who deny a problem exists, ascribing it (like the Determinism–Free Will problem) to a simple confusion. But before we turn to those, I want to say something about this intellectual endeavor called "philosophy".

The Philosophical Discontent

Every philosophizing starts from a sense of paradox.

(Hilary Putnam)

What is the subject matter of philosophy, and what is special about the way philosophy studies this subject matter? Philosophers claim ownership of this problem, and just as in all other problems, they do not reach consent. But one thing is commonly accepted: that philosophy is second order thought. It does not look at the world, but at the thought about the world. The Danish philosopher Søren Kierkegaard (1813–1855) used a nice parable to explain this. You walk along a street and in a shop window you see a sign: "Shoes are repaired here." You go in and ask to repair your shoes, only to find that this is not a cobbler's shop — it is the sign that is for sale. The moral is clear: in philosophy it appears as if we are speaking about the world, while in fact it is the concepts that are studied. Not only does philosophy deal with thought, says Kierkegaard, but it also does it *en passant,* to borrow a term from chess. It uses concepts and thereby studies them.

This is a beautiful definition, and in certain variations quite widespread. But more than it solves a problem, it generates one. Signs are also part of the world, no less so than shoes. Human thinking is part of the world, and can be studied like all other parts. In fact, we do this all the time. Anthropologists study the thinking of "primitive" tribes, historians of science investigate the scientific thought of ancient civilizations and psychologists study the way individuals think and feel. All these do not evoke the least feeling of being philosophical. So, what is left for philosophy to do? Why should anybody confuse signs and shoes?

Putnam's citation at the beginning of this chapter provides a clue. Philosophy starts from a feeling of discontent, a sense of confusion, an annoying little stone in your shoe that doesn't give you respite. More than that — the famous problems are not pebbles but big rocks. They are explicit paradoxes, each pointing at a contradiction arising from the most natural premises in the world. What characterizes philosophy, says Putnam, is not its object but the special way it looks at the object. It is not searching for facts and their explanations, but it tries to untie tangles. Philosophy starts from a sense of the ground disappearing beneath one's feet.

Where do these tangles come from? At least in some cases the answer is "circularity". But before we embark on an attempt to justify this proposition, let us listen to somebody who devoted to the problem all his adult life.

A Therapist of Thought

Ludwig Wittgenstein, easily the best-known philosopher of the 20th century, struggled with philosophy all his life. He claimed that he did it against his will, his only wish being to get rid of it. He had two answers to the question "What is philosophy?" one held before 1930 and the other after. The first view was the more extreme. Philosophy, so he claimed, is one big conceptual error. The reason is that its questions are meaningless. Only questions of fact are meaningful, and philosophical questions are not like that. The real role of the philosopher is to point out this meaninglessness, thereby getting rid of the problems. Having done this, Wittgenstein continues, the philosopher should pull the rug beneath his own feet, and realize that his own words are meaningless too.

If you identify agonized squirming here, you are right. This was Wittgenstein's bread and butter also in his personal life. He was born into one of the wealthiest European families of the time, but wealth did not bring with it happiness. I doubt that he experienced a single day of happiness. All his life he was an isolated and difficult man. Three of his brothers committed suicide and a fourth, more famous than Ludwig during their lifetime, lost an arm at the beginning of the First World War, did not relinquish his ambition to have a career as a pianist, and struggled throughout his life with this decision, in great misery. Ludwig himself was prey to suicidal urges for a long time. Adhering to the theories of another tormented thinker, Otto Weininger (1880–1903), he believed that there is no reason to hold onto life if you are not a genius. Weininger himself committed suicide in the house in which Beethoven died. It is said that Wittgenstein gave up the suicidal thoughts only after Russell recognized his genius. For some reason he did not sense the self-contradiction: a genius does not have to be told that he is such.

As already mentioned, Wittgenstein's career was divided into two parts, ten years separating between the two. In 1903 he left Vienna, his city of birth, to go to study aeronautical engineering in Manchester. There he met Russell's writings on the foundations of mathematics, and went to Cambridge to study under him. In the war he enlisted in the Austrian army and was taken prisoner by the Italians (then fighting for the side that they fought against in the Second World War). In the prison camp he managed to do what he never succeeded in doing at any later stage: compile his thoughts into a book. The book was later called the *Tractatus Logico-Philosophicus,* or *Tractatus* in short. The name was suggested by George Edward Moore, another Cambridge philosopher, in the footsteps of a book by Baruch Spinoza, the *Tractatus Theologico-Politicus.* The book, published in 1921, claimed to solve all philosophy's problems, and as explained above — a Gordian knot solution: declaring them all to be meaningless.

"Whereof one cannot speak, thereof one must be silent," Wittgenstein seals the *Tractatus.* And indeed, he kept silent for ten years. After the publication of the *Tractatus* Wittgenstein spent ten years in non-philosophical occupations. Five years he spent as a teacher in a small Alpine village, only to learn the limitations of his personality. He was harsh with the students as he was with himself. A student whom he beat tried to commit suicide, and he had to return to Vienna, escaping trial only barely, through the use of personal connections. He then tried his hand at architecture, designing his sister's house with the aid of a well-known architect. The house, not surprisingly, turned out austere and sharp angled.

All through this period his theories continued to reverberate in two places: in Vienna itself and in Cambridge. In Vienna a group naming itself the Viennese Circle, led by Moritz Schlick, met weekly to discuss his ideas. They tried to draw Wittgenstein to their meetings, but when he came he sat with his back to them and read aloud poems by the Indian poet Rabindranath Tagore, whose mysticism was meant to upset the rationalism bent group. In Cambridge it was Russell who spread the gospel. Eventually, in 1931, Russell and his friends succeeded in drawing him to Cambridge where he stayed almost until his death in 1951.

The center of gravity of the Analytic Movement, as Wittgenstein followers called themselves, moved to England. One reason was Nazism. Schlick himself was shot on the stairs of the University of Vienna building, by a student who mistook him for a Jew (and was later paroled, in consideration of his motive — after all, he meant well). Other philosophers emigrated to England and to the United States. Gilbert Ryle, an Oxford philosopher who assimilated Wittgenstein's second period ideas, took up the torch and applied these ideas to the Mind–Body problem. The book he published on the subject in 1949, *The Concept of Mind*, became a beacon for the discussion of the problem, at least for a while.

Ryle Visits a University

Ryle's idea was that mental concepts are constructed from physical ones. To clarify it, he used a story that became known as The University Parable.

> A man visits a university for the first time in his life. He spends a day on campus and is shown the buildings, the lecture halls, the students, the lawn, the occasional professor roaming the paths. At the end of the day he turns to his hosts and says: "All this is very nice. But where is the university?"

This is precisely what happens with the concept of "mind", so claims Ryle. We collect physical components and use them to form mental concepts. We see a person being pricked by a needle, hear him scream, see his hand recoiling, hear the words "it hurts" (which is a physical phenomenon — air being exhaled through vocal chords), and if we are neurologists we can study the signals transmitted in his nerve systems. From all this we form the concept of the "pain" of that person. But then something strange follows: we ask "but what is the pain itself?" Ryle called this error "a category mistake", but in fact such an inane way of thinking does not deserve such a respectable name. Looking at the wheels of a car, at its engine, seeing it ride, and then asking "but where is the car itself?" — is pure silliness.

So, the Mind–Body problem is just a result of confusion, says Ryle. The pain of a person is just his hand recoiling from fire, his scream, and the signals transmitted in his nerve system from the hand to the brain. These components **are** the pain — there is no ghost that is "pain" beyond them. They are not manifestations of some ethereal entity — they **are** the entity.

Another story Ryle uses to show the stupidity in looking for the "mind" beyond its physical manifestation is the parable of the locomotive and the horses.

> Peasants see a train for the first time in their life. They are accustomed to having horses perform such feats of mobilization, so they are sure that inside the engine of the train there are horses. The educated village pastor invites them to peer into the machine, and see for themselves that there are no horses there. They look in and indeed there is nothing there. Adhering to their theory they say, "Indeed, there is no horse. But there is a spirit of a horse." The pastor insists: but horses, too, are made of flesh and blood, and yet they move. This perplexes them for a while, it never occurred to them to ask how horses can move. "Yes," they eventually say, "inside the horse there is a spirit of a horse."

If Ryle is right, and the error behind the Mind–Body paradox is so simple, then it is not clear how the problem could persist for even one day, let alone 2400 years.

From the Inside or From the Outside?

Just as Hobbes, Hume and Mill failed to convince their fellow philosophers that the Determinism–Free Will conflict is imaginary, so did Ryle fail in finally resolving the Mind–Body problem. No philosopher put down his pen and said, "How did I miss this?" And just as in the problem of free will, they are right. Ryle's argument has a basic flaw: it does not explain why the "category mistake" occurs in the case of mental concepts and not in the case of universities and cars. What is so special about mental concepts?

Indeed, Ryle misses the crux of the matter. The arrow he shoots misses not only the center of the board but the board itself. He was right as far as mental concepts of others are concerned. We indeed form the concept of another person's pain from his physical reactions. We are not looking for a mysterious entity of the "pain" beyond his scream and the recoiling of his hand. But the heart of the problem is not in the concepts of other people's mental events. It is in the perception of one's **own** mental events. There Ryle's concept of mind is way off mark.

The essence of the Mind–Body problem is in that man's perception of his own mental events is totally different from his perception of other people's minds. Clearly, in order for me to know that I am in pain I don't have to hear myself say "it hurts" or watch my hand recoiling from the pricking pin. I don't have to see my irises contract (in fact, I cannot even see them contract) in order to know that I see light. I know it directly from the sensation of light itself. My thoughts are not only the words going through my mouth or keyboard, or the activity in my neurons. I know about my thoughts from their mere existence, without having to hear myself speaking them aloud. Or, at least, so it seems.

So, the Mind–Body problem is not at all about the gap between the material and the mental. It is about the gap between our acquaintance with other people's mental events and our own. One's own mind is a private world, accessible only to oneself. Here is what a person who devoted much of his career to the problem has to say about it:

> The main point is that things in the mental world are known with certainty only to one person, the owner of the mental event, and are not known to anybody else. They belong to the private domain, and from this point of view everybody has his own world.
>
> (Yeshayahu Leibowitz, *The Psycho-Physical Problem*)

This is a source of a paradoxical state of affairs. Other people's mental events can be mere physical reactions, but this cannot be the case for one's own mental events. If my pain is not merely my reactions, then it is indeed not physical. It belongs to another realm. It is here that the mystery starts: how come non-physical phenomena affect and are affected by physical events?

But look what this means: that the Mind–Body problem disappears once viewed from the outside, namely upon separation. It cannot even be formulated regarding another person, whom we observe from outside. As you will recall, this is clear indication that the problem originates in circularity. Let us see in what way.

Direct Knowledge

A thought can never be the object of itself.

(William James, American author and philosopher, 1842–1910)

If we do not know about our own pain from the recoiling hand and the scream, how do we know about it? The answer has already been given: by the pain itself. The pain is also its own knowledge. There is no need for the indirect mediation of other sensations in order to know that we feel pain or love, that we think, want, sense light. We know it by the mere existence of the pain, love, thought and will. Let us call this mode of knowledge "direct knowledge", and the assumption that this is the way we know about our own mental events the "assumption of direct knowledge". This assumption is the origin of the Mind–Body problem, since it means that there are no physical events in the world by which we learn about our mental events. And so, these events are left bereft of physical attributes. They cannot be viewed as part of the physical world, and the concepts regarding our own mental events remain devoid of physical objects.

Let us see what this assumption means in the case of light. Direct knowledge means that knowledge about the sensation of light is identical with the sensation of light. But this should mean that the light and its sensation are identical — there is nothing about the sensation of light that discerns it from the light. But, of course, the sensation of light is not the light, just as the stroking of a cat is not the cat. It is a relationship between light and the person sensing it. To know that "I sense light" I need to know at least that there is something called "I".

So, on one hand the concept "sensation of light" is denied the object of physical reactions, and on the other hand it is denied the substitute that naturally follows from this first denial — the light itself.

The following two diagrams may help clarify this. The first depicts the observation of another person's sensation of light.

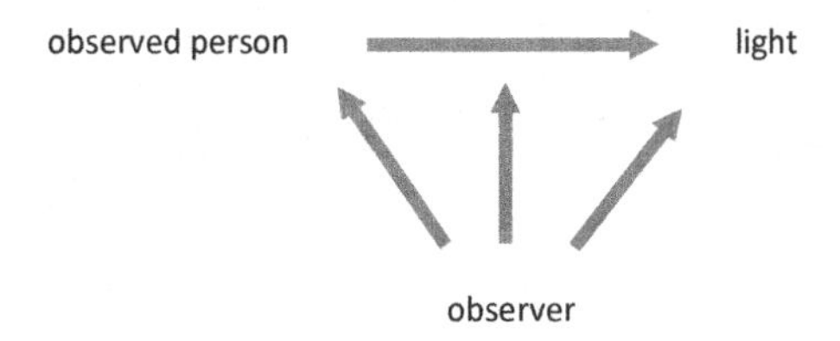

The top arrow denotes the sensation of light and the bottom ones the observation of the observed person, the observation of the sensation, and the observation of the light itself.

Self observation should be the same. When observing oneself, one should establish in himself a separate observing entity. This is not what happens in the direct knowledge assumption. There, observer and observed converge, as is signified by the double line on the left of the next diagram:

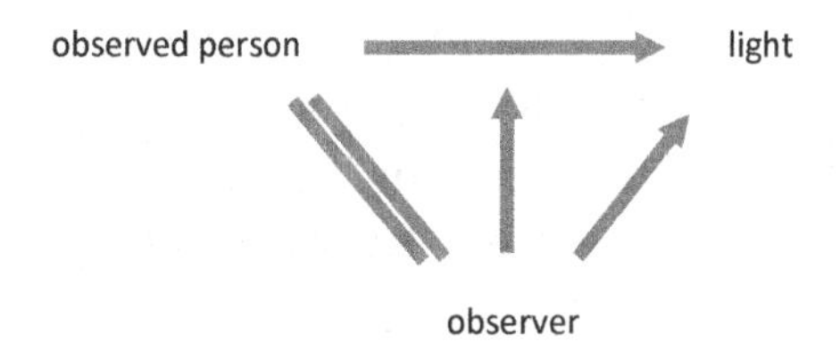

The identity of the observer and observed means that the right hand arrow is identical with the top arrow. And the direct knowledge assumption means that the middle arrow is identical to both: knowing about the sensation is identical to the sensation. But as noted, this leads to the impossible conclusion that the sensation of light is the light. It means that the concept of "sensation of light" is structured circularly.

As self-evident as the direct knowledge assumption seems, it is faulty. It leads to circularly defined concepts: a knowledge that is knowledge of itself, a sensation that is identical with its object. Let us see what two eloquent thinkers have to say about this idea.

Socrates and Lewis Carroll on Direct Knowledge

Socrates left nothing in writing. He always claimed that he didn't know anything and that his only advantage over others was that he was aware of that. Perhaps this wasn't a complete sham. He may have really meant it, and therefore did not presume that his thoughts deserve to be passed on to posterity. Plato, his student, was 29 when his mentor was executed for heresy and for corrupting the youth. This marked him for life. He did write a lot, mostly in the form of dialogues, in which Socrates is one of the participants, and in which he leads his interlocutor to the conclusions he wants. The early dialogues were probably written from memory, and it is commonly assumed that they indeed reflect Socrates' theories. Afterwards he used the medium he created to express his own opinions.

Charmides is one of the early dialogues. Socrates lucidly expresses in it the problematics in the direct knowledge assumption. Kritias, the other participant in the conversation, suggests at some point that "sobriety" is "the knowledge of itself". Detecting an inaccuracy, Socrates hangs on to his prey and does not let go.

> But look, my friends, how strange it is what we are trying to say. Do you think that there can be such a thing as a sight that is not the sight of anything in the world, but is the sight of its own self, and of other sights? ... And how about hearing, that is not of a voice, but of itself? ... Do you believe that there exists a sense that is the sense of itself and of the other senses, but does not perceive

> anything of the sort that other senses perceive? ... And a will that does not strive for anything good, but wants itself and other wills?

All these poor Kritias has to answer in the negative. There is no such thing as "knowledge of its very self". Socrates is speaking precisely about the identification of mental events with their perception. "Knowledge that is the knowledge of itself" is knowledge that needs no observation in order for us to know it is there.

Lewis Carroll had his own go at the assumption of direct knowledge. In *Through the Looking Glass* there is a dialogue that is a parody on this assumption: it is a succession of confusions between objects and their names, namely between objects and relating to them. The old knight (who is also a chess piece) tells Alice that since she is sad, he is going to recite to her a song that is so sad that it fills the eyes of everybody hearing it with tears, or else ... . "Or else what?" asks Alice. "Or else it doesn't."

> "... The name of the song is called '*Haddocks' Eyes*'."
>
> "Oh, that's the name of the song, is it?" Alice said, trying to feel interested.
>
> "No, you don't understand," the Knight said, looking a little vexed. "That is what the name is *called.* The name really is '*The Aged Aged Man*'."
>
> "Then I oughtn't to have said 'That's what the *song* is called'?" Alice corrected herself.
>
> "No, you oughtn't: that's quite another thing! The *song* is called '*Ways And Means*': but that's only what it's *called,* you know!"
>
> "Well, what *is* the song then?" said Alice, who was by this time completely bewildered.
>
> "I was coming to that," the Knight said. "The song really is '*A-sitting On A Gate*': and the tune's my own invention."

Just like in the confusion between the pain and its knowledge, or the sensation of light and its knowledge, there is here confusion between first and second order. And, as befits a parody, this confusion goes on to third and fourth orders.

How do People Learn about Their Mental Events?

How can I know what I think till I see what I say?

(E. M. Forster)

The Mind–Body problem is born from circularly constructed concepts of our own minds, lacking separation between observer and observed. The cross one should therefore hold against it is separation of powers: between first and second order, between examiner and examined, between a participant in the game and the spectator. The obvious must be accepted: that a person knows himself in the very same way as he knows others, at least in principle. To know about his mind he has to observe his reactions, just as he does when he learns about other people's minds.

Why don't philosophers accept this, and why do they hold onto the direct knowledge assumption for dear life? After all, everybody knows that playing football and watching the game are different things. How come philosophers accept so willingly the assumption that knowing about the sensation of light is the same as the sensation of light? Why is it that when it comes to one's own mind, first order and second order are so easily confused?

The reason is historical. Not the history of mankind but the individual history of each of us. After the concept of "I" is formed, we learn to make a shortcut from "light" to "I see light". Tautologically, if I say "here is light" I can deduce "I see light", and with the years we learn to automatically deduce the second from the first. We forget that these are two different things, and that eons ago we did not know at all that there is an "I". We had to construct this concept with some effort. In fact, the concept of the "I" is probably first constructed from other people's actions, mainly our mothers relating to us, which teaches us that "I am here".

The conclusion is strange, but inevitable. Indeed, a man has to watch his reactions in order to learn about his mental events. He has to hear himself screaming in order to know that he is in pain, and to watch himself reacting to the light in order to know that he sees light. He has to hear himself saying "I want this" in order to know about his will. Without knowing that there is such a thing as "he himself", the feeling of pain is just "pain", not "I feel pain", the sensation of light is just "light", and the wish to have "this" is just "This! This!", that young children say.

The Cat Sits, Therefore the Cat Is

On a cold night in November 1629 the philosopher-mathematician René Descartes, then a soldier of some small German duchy, was sitting in an isolated cabin in the snow fields of Bavaria. Loneliness suddenly enveloping him, he was gripped by terrible anxiety. Isn't it possible that his senses are misleading him? That some wicked demon is playing a jest on him, manipulating his sensations so as to make him believe in certain things that have no link to reality? For example, couldn't the demon direct the light beams that reach his eyes, or make him hear certain sounds, so as to cause him to believe they arrive from the external world? How can he be sure that there exists an external world at all? His consciousness is the prisoner of his perceptions. There is no route from the world to the brain that circumvents them. Once he doubts his perceptions, there is no firm ground on which to soothe the doubts, no way to refute them.

The appropriate name for this fear is "circularity anxiety". Descartes wasn't the first to succumb to it — hardly any philosopher was free from it. But unlike his predecessors, Descartes swore not to give in to it, and not to rest until he finds firm ground, an Archimedean leverage point that cannot be doubted. He believed that he found one. When in doubt, one certainty remains: that he is doubting. And if he doubts, then he thinks. But if he is thinking, then in particular he **is**. "I think therefore I am." This was the cross he held against the ghost of doubt and it became the best-known philosophical dictum.

But why did Descartes decide that of all things the fact that he thinks cannot be doubted? What is the difference between "I think therefore I am" and "The cat sits, therefore the cat is"? ("The cat sits on the mat" is a favorite example with philosophers.) Why is the first a glorious response to the doubt and not so the second? Descartes' answer is that knowledge about the sitting of the cat necessitates the mediation of our sensations, whereas the fact that I think does not. I know it directly. This is the reason that "Joe thinks therefore Joe is" doesn't work (again, by Descartes' argument). About Joe's thinking I do not know directly, I can only assume it from Joe's physical reactions.

"I think therefore I am" is then but another expression of the direct knowledge assumption. This explains why many consider this dictum to be the origin of Western philosophy, and think that modern philosophy sprang from Descartes' forehead on that fateful night. It connects with Popper's view that Western philosophy revolves around the Mind–Body problem. Both are based on the belief that knowledge of our own minds is totally different from the knowledge of any other part of reality. This is the belief that prevents philosophers from studying signs without confusing between them and their contents — the confusion that Kierkegaard talked about in his definition of "philosophy".

The Red King's Dream

If a clown should say that his bailiff does not exist though he stands in front of his very eyes, he would be taken for a madman, and for good reason. But when a philosopher says the same, he expects us to admire his knowledge and his sagacity, which infinitely surpasses common apprehensions.

(Leonhard Euler, in *Letters to a German Princess*)

Descartes' circularity anxiety was an expression of an approach that is often identified with philosophy: skepticism. "The theory of knowledge is nothing but an exercise in skepticism," said the English philosopher Alfred Ayer. Skepticism advances the idea that we are prisoners of our sensations. They are the only thing we possess direct knowledge of and there is no way for us to know that there is anything beyond them.

Skepticism is based on a circular perception of ourselves. Evidence of this is that it disappears with separation. One can only apply it to oneself. There is no point in being skeptical about somebody else's perceptions. If another person testifies that he sees a tree, all you have to do if you suspect him of imagining this tree is look for yourself. If there is one, the person is not dreaming. This is ordinary, non-philosophical doubt. However, if doubt is cast on your own testimony that there is a tree there, looking for the tree will not help you. Your second testimony will be useless, because it is precisely that testimony that was doubted. Substantiating a testimony on the grounds of itself is circular.

This is what Euler is trying to convey in the citation above. Lewis Carroll entertained no better opinion of the skeptical argument, and he devotes to it one of his sharpest parodies on philosophy, with which *Alice in Wonderland* and *Through the Looking Glass* are strewn. In *Through the Looking Glass* Twiddle Dee tries to convince Alice that she is merely a figment of the Red King's dream.

> "And if he left off dreaming about you, where do you suppose you'd be?"
>
> "Where I am now, of course," said Alice.
>
> "Not you!" Twiddle Dee retorted contemptuously. "You'd be nowhere!"

This is a parody on Bishop Berkeley's assertion that we are only figures in the dream of God. It is of course also a mirror of a mirror, for *Through the Looking Glass* is the story of a dream of Alice's, so in fact Alice is dreaming the Red King, who is dreaming her — circularity epitomized.

Carroll is taking the bull of the skeptical argument by the horns. If the problem is senseless from aside, he is telling us tongue in cheek, it is as ridiculous when formulated from the inside.

The Illuminated Side — Scientific Breakthroughs

Part IV: Large Infinities and Still Larger Ones

Infinity is the place where things happen that don't.

(Anonymous student)

You Shall See the Land from Afar

So far we have met circularity from its inimical side, that which confuses people and diverts them from the straight and narrow. But just as historical figures have many personae (what was Napoleon — a big criminal or a great man?) so does circularity have another facet, which has proved of great value to science. It emerges when phenomena of circularity are unveiled, usually thereby supplying proof to the impossibility of some task. In the final analysis, the scales weighing the virtue and vice of circularity would tilt towards the illuminated side. Going around in circles means not progressing, and therefore the dark side of circularity hasn't changed much in the last two millennia. The illuminated side, on the other hand, has partaken in the progress of scientific activity.

In this part we shall meet one discovery based on circularity, a mathematical theory developed by the German mathematician Georg Cantor (1845–1918). Cantor worked in the northern German town of Halle, whose main claim to fame is as the birth place of another Georg — the composer Georg Friedrich Handel. Cantor used a circularity argument to prove a "non-existence" result — that of the largest set in the world. It does not exist. Every set has one larger than it.

A profound discovery entitles the discoverer to a place in the scientific pantheon, and Cantor did indeed win this well-deserved glory, unfortunately mainly posthumously. In his lifetime the discovery brought him mainly heartbreak. When he set on his trail of research he sought neither the glory nor the struggle attending the discovery. Having started his career in a classical area that deals with the shape of waves, he had made his discovery almost against his will. At some point in his research he noticed a surprising fact that ultimately proved much more important than the original line of research: that there is more than one type of infinity. There are large infinities and there are still larger ones.

Until Cantor's time mathematicians distinguished between finite sets, that are "small" (compared to infinity), and infinite sets, that are "large". Of course not all finite sets are equal in size. A set of 3 is larger than a set of 2. But with regard to infinite sets it was assumed that such a distinction does not apply, that they are all "large" and that's it. Cantor discovered that in the kingdom of infinite sets there are subtleties too, and that just as a finite set can be larger than another so can an infinite set be larger than another.

This insight generated a new field — set theory, which investigates some of the most fundamental mathematical concepts. So much so, that today set theory is taught in most universities in first year. Cantor's bad luck was his timing. He made his discovery precisely when other mathematicians reached conclusions that at least superficially appeared to be contrary to his own. Cantor spoke of infinity as an existing entity at a time his colleagues referred to it disrespectfully as "actual infinity", arguing that there is no real infinity. Infinity is not achieved, it can only be approached by numbers that are larger and larger.

It all began when the 19th century mathematicians tried to put some order in a field that was already two hundred years old then: differential and integral calculus (often called just "calculus"). This is the mathematical field that studies change, such as we find in motion — the change of place over time. You may wonder — isn't motion studied in elementary school? "Car A goes at a speed of 50 kilometers an hour. If car B leaves from the same spot an hour later, at a speed of 100 kilometers an hour, in the same direction, when will it catch up?" The point is that in elementary school cars go at constant speed. In real life this rarely happens. Bodies change their speed. So, to measure speed it does not suffice to say "the speed of a car travelling 100 kilometers in 2 hours is 50 kilometers an hour." You'd then be speaking of its average speed. To calculate the real speed (even to define what it is) you need to measure motion over smaller and smaller time intervals. The smaller the interval, the more accurate is the approximation of the real speed. In other words, you should look at time intervals that tend to zero.

The mathematicians of the 17th and 18th centuries spoke of "infinitely small intervals", a term they never entirely clarified. In the 19th century it became clear that such fuzzy concepts cannot be used for rigorous proofs. There is no such thing as an "infinitely small" quantity. Instead, another notion was invented: "as small as we please". Instead of one "infinitely small" number they spoke of a sequence of numbers that goes to zero. For example, the sequence $\frac{1}{2}, \frac{1}{3}, \frac{1}{4}, \ldots$ tends to 0, meaning that its terms get as close to zero as we wish.

Similarly, on the other side of the spectrum, where numbers were until then "infinitely large", they became "numbers as large as we please". For example, in the sequence 1, 2, 3, ... the numbers get as large as we wish. Do you want the terms to be larger than a million? No problem: as from the million and one place all terms are larger than a million. In this formulation there is no actual infinity. "... *Thou shalt see the land before thee; but thou shalt not go thither,*" as God told Moses when he approached the land of Canaan (*Deuteronomy*, 32:52).

In Cantor's world the Promised Land is reached. He speaks of sets that are truly infinite. This obviously aroused strong resistance. To rock the foundations of mathematics that have been so recently and so painstakingly erected, and to reinstate within the fort the newly exorcised demon of actual infinity, was heresy or worse — lack of understanding. Henri Poincaré, a leading mathematician of the time, called set theory "a childhood disease from which mathematics will recover in due course". Cantor's most bitter enemy was Leopold Kronecker, an influential member of the German academy. Kronecker called him a "mathematical charlatan", maintained that "there may be philosophy or theology in Cantor's work, but certainly no mathematics", and accused him of corrupting the youth, the charge for which Socrates paid with his life some 2300 years earlier. As a result Cantor failed to get a much desired university position in one of the leading universities. This and the tragic loss of a son in 1899 aggravated a depressive illness from which he suffered, and he ended his life in a mental asylum.

What is "Counting"

Count them.
You can count them. They
Are not like the sand on the seashore. They
Are not innumerable like the stars. They are like lonely people
On the corner or in the street.

(Yehuda Amichai, *In the Full Severity of Mercy*, 1st stanza)

"The size of an infinite set" is seemingly a complex notion. By contrast, "the size of a finite set" is simple: isn't it just the number of its elements? In fact it is not simple at all. Have you ever contemplated the question of what is a number? This is such a basic notion that we never stop to examine it. As far as we are concerned, a set of three apples comes to the world complete with the label "3". Well, not quite. Numbers are a tool invented by man, one of his most sophisticated.

To understand the wisdom of numbers, think of a person who wishes to compare the height of two tables in his apartment: the dining table and his desk. One way to do it is to haul the table from the study to the kitchen (or vice versa), stand them side by side and find which is higher. A simpler way is to take a stick, mark on it the height of the desk, carry the stick to the kitchen, and compare the height of the notch with the height of the dining table.

Numbers are such a "stick". In order to know if a zoo population comprises more giraffes or more elephants, we do not need to put them side by side. Instead, we can employ as a "measuring yard" a standard set — that of the numbers 0, 1, 2, 3, … . Like the stick, this is a mediator. Counting the elephants means assigning to each an element from this set, in order. We then proceed to the giraffes' compound, count them, and measure on the "stick" of numbers which notch is higher. This stick is even better than the material one, because it can be carried in the head.

It is quite incredible, but for many tens of millennia nobody stopped to wonder what they are doing when they count. For Cantor, wishing as he did to measure infinite sets, this was an insight he could not go without. In the infinite case, the direct method (i.e. without the mediation of numbers) is the only way. Ordinary numbers are far behind us and "infinity" is not a number. Hence, in order to compare infinite sets, we must "put them one against the other", namely establish direct correspondence. This was the crucial observation that led Cantor to his discoveries. The following is what defines two sets as equal in size:

> Two sets have equal size if they can be matched, one element of the first against each element of the second.

For example, if you want to show that both your hands take pride in the same number of fingers (if luckily this is the case) you do not need the mediation of the set of numbers {1, 2, 3, 4, 5}. You can simply put your hands one against the other, finger against finger, thereby forming a one-to-one correspondence. The arrows in the following diagram show that the set of flowers is equal in size to the set of pots.

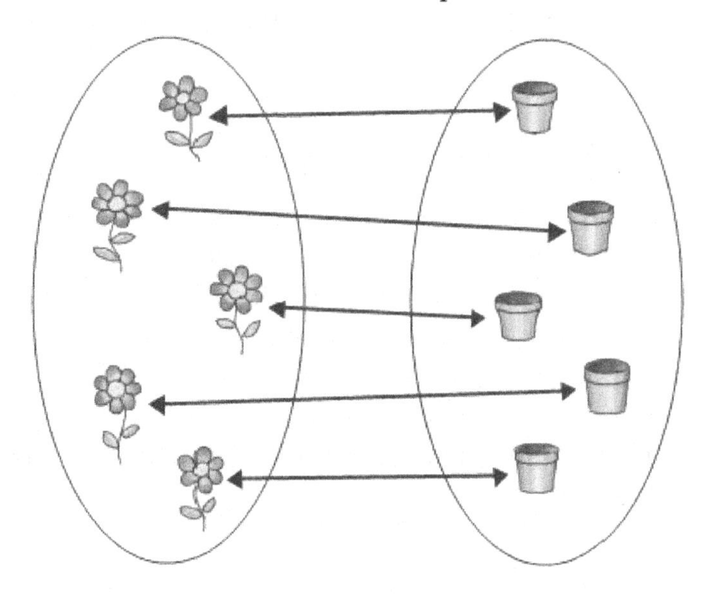

A Surprise

You don't understand mathematics. You just get used to it.

(John von Neumann, mathematician)

What are there more of in the world — humans or women? Humans of course. After all, women are only part of humanity. There are also men. The set of women is just part of the set of all human beings, and so it is smaller. There are about twice as many humans as women.

What are there more of in the world, natural numbers or even numbers? Natural numbers of course. After all, even numbers are only part of all numbers. There are also odd numbers. The set of even numbers is just a part of the set of all numbers, and so it is smaller. There are twice as many natural numbers as even numbers.

True, or not? Only partially. Indeed, there are twice as many natural numbers as even numbers, but in the infinite case "twice as many" doesn't mean "more than". Surprisingly, the two sets are equal in size. This can be shown by matching, just as you can show that both your hands have the same number of digits, by putting them one against the other. We can assign to every natural number an even number, in the following way. To zero we assign 0; to 1 we assign 2; to 3 we assign 6, and so on: to every number we assign the number multiplied by 2. This forms a matching in which to every number an even number is assigned, and every even number is assigned to precisely one number — half of itself. The matching is depicted in the following diagram:

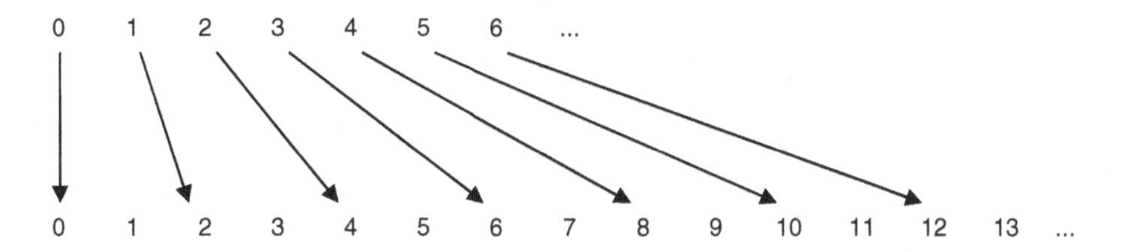

I deliberately listed the odd numbers as well (bottom row), so that it is clear that the top set is matched to what is only a partial set of itself.

The story of Tristram Shandy, the protagonist of Laurence Sterne's book, *The Life and Opinions of Tristram Shandy, Gentleman*, illustrates this principle. Shandy is busy writing his autobiography. He is so fastidious, that it takes him a hundred days to describe each day of his life (in the book it is a bit different, but this is the idea). Will he ever complete his task? If his lifespan is finite then of course not, but if he lives forever he will. For example, on the 100,000th day of writing he will be recording the 1000th day of his life. Tristram Shandy has one more reason to wish to live forever than the rest of us.

Cantor's set theory had a guardian angel in the form of David Hilbert (1862–1943), the most prominent mathematician of the late 19th and early 20th centuries. He fell in love with the new theory and became its proponent. "From the paradise that Cantor created for us nobody can expel us," he said. One of his contributions was a story illustrating the special features of the concept of infinity. The hero of the story is the owner of a hotel, somewhere in heaven (perhaps in Cantor's paradise), that has infinitely many rooms, numbered 1, 2, 3, ... One day the hotel was full to capacity, occupied to its last room (well, it doesn't really have a last room). And then, in the evening, another guest came, gasping for breath. Had the hotel been finite he would have had to sleep at large. Not so in an infinite hotel. The manager used an infinitely powerful loudspeaker to ask his guests each to move to the next room. So, the guest in room number 1 moved to room number 2, the guest in room number 2 to room number 3 and so forth. Every guest again had his own room, and one room had become vacant — room number 1, to house the new guest.

Then, again, by the evening of the next day all rooms were occupied, when something even more disastrous happened: suddenly infinitely many new guests arrived. Again the manager did not lose his presence of mind. He told every guest to move to the room bearing the number double to his present one. Namely, the guest in room number 1 moved to room number 2, the guest in room number 2 to room number 4, the guest in room number 3 to room number 6 and so on. Every guest had a room of his own and infinitely many rooms were vacated: all the odd-numbered ones. Room was found for the infinitely many newcomers.

Inequality Between Sets

All animals are equal but some animals are more equal than others.

(George Orwell, *Animal Farm*)

Infinity holds more and more surprises. A set that looks much larger than another may turn out to be of the same size. For example, it seems that there are many more fractions than natural numbers. After all, between every two natural numbers there are infinitely many fractions. And yet, the set of fractions is equal in size to the set of natural numbers.

How do you show it? Again, by a matching, namely assigning a fraction to every natural number, in such a way that all fractions appear in this matching. A little thought reveals that such a matching is nothing but counting all fractions: the fraction assigned to 1 is the first, the fraction assigned to 2 is the second, and so on. So, such a matching means that the set of fractions is "countable", namely it can be counted.

Let us first allow some concession: let us count only the non-negative fractions. Here is the trick that does it. We arrange the non-negative fractions in order of the sum of their numerator and denominator. The fraction with the smallest sum is $\frac{1}{1}$, in which the sum is $1 + 1 = 2$. This is the first fraction in the counting. Then come the fractions with 3 as the sum of their numerator and denominator. These are $\frac{1}{2}$ and $\frac{2}{1}$. We put them as second and third in the list. Then come the fractions with sum of numerator and denominator 4. These are $\frac{1}{3}$, $\frac{2}{2}$ and $\frac{3}{1}$. We count them as numbers 4, 5 and 6 in the list (in fact we could skip $\frac{2}{2}$, because it is equal to $\frac{1}{1}$, that we have already counted, both being equal to 1).

A slight variation will enable us to count all fractions, including the negative ones. The idea is to go alternately — taking a positive fraction, then a negative one, then a positive one and so on.

Another surprising fact: it seems that on the infinite line there are infinitely many times as many points as on a finite segment. But in the infinite case, "infinitely many times" does not mean "more". The correspondence in the picture below shows that the number of points on a segment (without endpoints) is equal to the number of points on the entire line. Bend the segment to form a semi-circle and put a source of light at the center of the circle. The light rays would then form a one-to-one correspondence between the points on the segment and the points on the line, their "shadows".

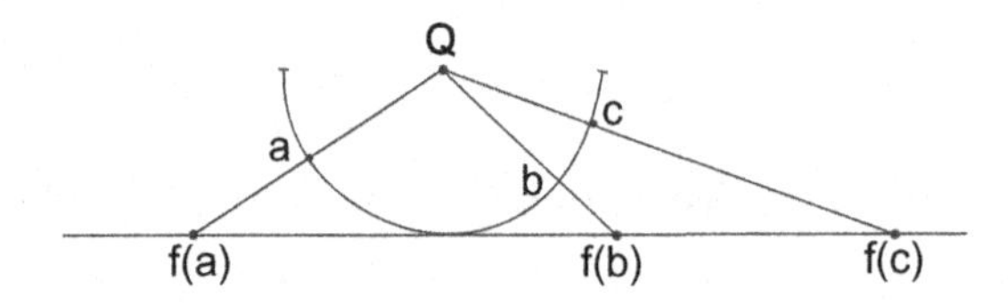

The rays emanating from Q match the points on the semi-circle with those of the infinite line.

Judging by these examples, the suspicion may arise that all infinite sets are of equal size. In order to know whether this is the case, we should first define when one set is bigger than another. Following Cantor, let us first define when one set is larger than **or equal to** another set. Here is an example that will help us reach the definition. Look at the following drawing:

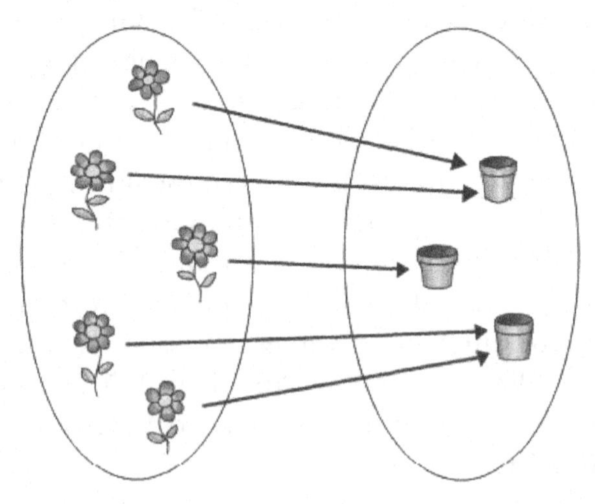

The arrows show that there are at least as many flowers as pots: they assign to each flower a pot, in such a way that all pots are assigned to at least one flower. There are sufficiently many flowers to "cover" all pots. If in a dancing club you wish to show that the number of women is at least as large as the number of men, just assign a woman to every man, not necessarily monogamously, in such a way that all men are matched. This means that there are enough women to "cover" all men.

We now know when a set is larger than another or equal to it in size. And when is a set A strictly larger than a set B? The answer is simple: when A is larger than or equal to B, but not vice versa, namely B is not larger than or equal to A.

In the terminology of correspondences:

A is larger than B if there is a correspondence from A to B that covers all of B, but the converse is not true.

With this definition at hand we can prove Cantor's theorem:

There is no largest set in the world.

In other words, for every set we can find a larger one. For the proof, let us play a famous children's game.

Cantor Plays with Mr. Potato Head

The popular toy Mr. Potato Head consists of a head to which all kinds of features can be added: a moustache, ears, a hat, a nose. In the original game each feature has many variations: the hair could have been blond, red or black, the moustache big or small. In our version we shall consider only two options for each item: to be put on or not. The figure either has a moustache or not, either has a left ear or not and so on.

This game was invented in 1949, thirty one years after the death of Cantor. But had Cantor known it, he could have used it to illustrate his theorem. In terms of the toy, Cantor's theorem says this:

There are more characters than features.

Cantor proved that the number of figures you can form with given features is larger than the number of features, even if the number of features is infinite.

For example, if there are no features at all, then the number of features is 0, while the number of possible figures is 1: the head with no features at all. And 1 is more than 0. If there is 1 feature, say a nose, then there are 2 figures: one with the nose and one without it. And 2 is more than 1. If there are 2 features, say a nose and a moustache, there are 4 figures: with both items, with none, with nose but no moustache and with moustache but no nose. And again, the number of figures, 4, is more than the number of items, which is 2.

If there are finitely many features, every feature multiplies the number of possible figures as each figure can "have or not have" this feature. So, the number of figures is 2 to the power of the number of features. If there are n features there are 2^n figures, and 2^n is larger than n.

How do you prove that there are more characters than features? Let us assume that this is not the case, namely that there are no more characters than features. We shall derive a contradiction from this assumption which will show that this is impossible. Remember what this assumption means? That the characters can be "covered" by the features. Namely, that there is an assignment of a character to each feature, so that all characters are "covered", i.e. matched to some feature. So, suppose that some salesman of Mr. Potato Head promises that he can construct such an assignment. Cantor will prove to you that the man is lying. There is no such assignment. He will do it by constructing, before our very eyes, a character that is not covered, contrary to the promise of the salesman. And for the purpose of this construction he will invoke circularity.

Let us call the character that Cantor builds "Mr. Spite", we shall soon see why. Cantor assembles it, feature by feature. For every feature Cantor is going to tell us whether Mr. Spite has it or not. Take, for example, the nose. Remember? The salesman assigned to it a character (he assigned to every feature a character). This character may have a nose, and it may not (the fact that he was assigned to the nose does not mean anything with regard to his actually having one). Regarding the nose Mr. Spite does the opposite of what this character does: he puts a nose if this character doesn't have one and will not put one if the character does have a nose.

Note that Mr. Spite cannot be equal to the character assigned to the nose: they differ on the nose. And it is enough for two characters to differ in one feature in order not to be identical.

Let us now turn to the moustache. The salesman assigned some character to the moustache (remember, again, that he assigned a character to every feature). This character may have a moustache and he may not. If he has, Mr. Spite will choose not to have a moustache. If the character assigned to the moustache does not have a moustache, Mr. Spite will choose to have one. Note that Mr. Spite is different from the character assigned to the moustache — they differ on the moustache.

It is now clear how Mr. Spite fully deserves his name: for every feature, he chooses contrary to the character assigned to that feature. He differs on the hair from the character assigned to the hair, and on the left ear from the character assigned to the left ear. The result? Mr. Spite is different from the character assigned to **any** feature. What this means is that he is not assigned to any feature. This argument works for any assignment. Every assignment has its own "Mr. Spite", a character that it misses. There are plainly too many characters to be covered. Which means that there are more characters than features.

From here the way is short to Cantor's theorem:

For every set there is a set larger than it.

How do we prove this? All we have to do is realize that every set can be the set of features. For example, a Potato Head with mathematical inclinations can use as features the natural numbers. On Monday he can wear all even numbers, and on Tuesday all prime numbers. No place on his figure to put them all? Let him use his imagination.

So, we can persuade our Potato Head to use any set S as a set of features. Cantor's theorem says that whatever the set S is, the set of characters is larger than S. So, every set S has a set larger than it — the set of characters, if the elements of S are used for features.

Finally, a mathematically minded Mr. Potato head will also explain to you what a "character" is: it is just a set of features. Clearly, it is the set of features that determines what the character is. In other words, the set of characters is the set of partial sets (also called "subsets") of the features set. And since every set can serve as a set of features, Cantor's theorem actually reads:

The set of subsets of a given set is larger than the set itself.

In *Part VII For the Experienced Hikers,* I will tell you the original form of Cantor's proof, why it is called the "diagonal method", and why the theorem entails that there are more real numbers than natural numbers.

The Paradoxes of Set Theory

The definition of Mr. Spite feels circular. Look what features Mr. Spite chooses: precisely those that the character corresponding to them does not put on. Whether or not Mr. Spite puts a feature on is defined by the character corresponding to the feature. I almost said "by the feature itself". With a little distortion, this "almost" can be made to be "really". The definition of Mr. Spite is not circular, but with a bit of cheating and camouflaging, it can be made to be so. If this be done, Cantor's theorem would become a paradox. Some people have actually done precisely that. It has happened in the history of mathematics not once but four times. I will tell you of two of the four paradoxes that were thus generated, those of Russell and of Cantor himself.

Chronologically, the first was Cantor's paradox. We know that for every set there is a set larger than it. But can this be true? Consider, for example:

The set of all things in the world.

By "things" we mean also abstract things, like sets. Since this set contains everything, it is the largest set in the world. No set can be larger than it. But this contradicts Cantor's theorem, which says that every set has a set larger than it!

Cantor was not particularly perturbed by this paradox, or by another he found, that was rediscovered by Cesare Burali-Forti a few years later and eventually named after the latter. With prophetic intuition he sensed that they would be easy to rectify, and would not harm his theory. Somebody else however did get worried — Bertrand Russell. Russell was not a creative mathematician, but had a sound mathematical education, and knew what to do with a contradiction when it stared him in the face: try to track its footsteps. If Cantor's theorem leads to a contradiction in the case of the set of all things in the world, then its proof should be analyzed in the special case of this set. This Russell did, and he discovered that the set "Mr. Spite" that emerges in this case is particularly simple. It is:

R: The set of sets that do not belong to themselves.

Most sets you can come up with are in R. For example, the set of chairs is in R, since it does not belong to itself — it is not a chair itself. But the set of all sets is a set itself, so it belongs to itself. Namely, it does not belong to R.

Russell's paradox appears when you ask a simple question: does R belong to itself? By its definition, this happens if and only if it does not belong to itself. But careful reading of the last sentence reveals a contradiction: R belongs to itself if and only if it does not belong to itself!

Russell formulated a nice parable to illustrate his paradox. In a small village there lived a barber, who had vowed to give a haircut to precisely those villagers who did not cut their own hair. But now he is in a quandary: should he cut his own hair, or not? If he gives himself a haircut, then, according to the vow he took, he cannot cut his own hair. But, if he will not give himself a haircut, then to be true to his vow he must!

To cut or not to cut?

The story of the barber is not a paradox. His vow simply cannot be realized. Russell's set, however, seemed to lead to a real contradiction. It is not a matter of a vow that cannot be fulfilled — R indeed exists. As mentioned above, Russell's paradox was nothing but a reformulation of Cantor's paradox, but because of its simpler formulation it had much greater impact. For a brief moment, the foundations of mathematics shook. The paradoxes threatened to expel us from Cantor's paradise.

The all-clear siren went off not long after. The mathematicians of the first decade of the 20th century soon discovered where the cheating was. It was in an innocent looking assumption called "comprehension". It is that for every property there is the set of all things satisfying this property. For example, the property "being a chair" generates the set of all chairs. Cantor and Russell showed that this assumption is not that innocuous. It can generate circular definitions. For example, if we assume that the property of sets of not belonging to themselves defines a set R, something is defined circularly — the belonging of R to itself. Comprehension contains therefore a mistake, and with regret the mathematicians of the beginning of the 20th century realized they would have to relinquish it.

Having had to give up the main tool for constructing sets, substitutes had to be found. How can we construct sets without risk of circularity? The German mathematician Ernst Zermelo suggested a way of constructing sets bottom to top. You start from the empty set, that does not have any elements, then you construct a set having one element (possibly the empty set itself), and then by some elementary operations like the union of two sets you construct larger and larger sets. This evades circular definitions, and vouchsafes set theory from re-appearing paradoxes. Peace of mind was regained. This brief tumultuous period in the story of set theory testifies once again to the value of paradoxes. Due to their discovery set theory nowadays seems to rest on sound foundations.

Part V: Gödel's Incompleteness Theorem

I am human and nothing human is alien to me.

(Terence (Terentius), a Roman playwright)

I am human, and nothing human is alien to me.
But also not particularly close.

(Dalia Ravikovitch, *Human Attributes*)

A Revolution in a Small Town

Twenty four is a good age for scientific revolutions. Isaac Newton was 24 when during one summer he developed the infinitesimal calculus and the theory of gravity, Albert Einstein was that age in his *annus mirabilis* 1905, in which he developed the theory of relativity and set the foundations of quantum theory, and Werner Heisenberg was 25 when he discovered the principle of uncertainty (which, by the way, has a lot to do with circularity). This list was extended in 1930 by the Austrian mathematician Kurt Gödel. He instigated a revolution that went far beyond the boundaries of its proclaimed field of mathematical logic. It had two seemingly divergent aspects. One can be summarized in **not every true fact can be formally proved**, and the second can be formulated as **there is no all-knowing computer program**. No single computer program can resolve every problem it tackles.

Both statements relate to numbers. The "true facts that cannot be formally proved" are facts about numbers. For example, "there are infinitely many prime numbers" is a true fact about natural numbers, provable using the customary systems of axioms for numbers. Gödel showed that every "reasonable" (in a sense to be explained below) system of axioms misses some true statements, meaning that it is incapable of proving these statements. This is Gödel's Incompleteness Theorem, and when people speak about "Gödel's theorem" this is what they usually refer to.

The second statement is less famous than the Incompleteness Theorem but is no less important, since it set the foundations of theoretical computer science. It says that there is no computer program that can settle the truth or falsity of every single statement on numbers with which it is challenged. The two statements are linked through their proofs: the second result is proved using the first. And more importantly for us — they both stem from the phenomena of circularity. Gödel's theory is the greatest victory of circularity ever.

Gödel presented his discovery in October 1930, on the last day of a conference held in the north German town of Königsberg. The year before he had proved a theorem that was at the time the most significant result in mathematical logic, called The Completeness Theorem. Don't worry — he did not prove two contradictory facts. The Completeness Theorem relates to the completeness of the Frege–Russell–Whitehead system of proofs, while the "incompleteness" is that of the Peano system of axioms. These are two different things: a proof system doesn't tell you what the axioms are, but how they should be applied. In *Part VII For the Experienced Hikers,* I will explain the difference between the two. But at the time few were aware of the significance of the Completeness Theorem, and not many heeded the shy and slight of build young man. Luckily the talk was attended by one of the brightest mathematicians of the 20th century — John von Neumann. Only three years Gödel's senior, von Neumann saw through the technical difficulties of the new theory, and became a vociferous advocate. Thanks to his intervention, news of the revolution quickly spread.

To understand what Gödel's revolution is about, we have to retrace about 80 years in mathematical history, to the point when its seeds were sown. It was an insight that changed our view of "thinking" — the realization that machines can also think.

Copernican Revolutions

A baby views itself as the center of the universe. Everything revolves around it. Maturation means, among other things, renouncing this position and recognizing the separate existence of others. Science has undergone a similar process. Primitive thought is anthropocentric: Man casts God in his mold and attributes natural phenomena to his own deeds. "If horses could sculpt," said the philosopher Xenophanes, a contemporary of Pythagoras, "they would form their gods in their template." Departure from this mode of thought was slow and gradual: science enjoys the privilege over an individual, of evolving over generations. But even in this gradual process some quantum jumps can be detected, for which Sigmund Freud coined the name "Copernican revolutions".

Freud counted three such revolutions. The first was the cosmological revolution of Nicolaus Copernicus himself, who understood that earth is not the center of the universe in any possible sense of the word. The second was Charles Darwin's, who taught us that though the most intelligent member of the animal kingdom, man is subject to the same laws that govern the kingdom's other species. The third revolution, Freud — who never suffered from too much modesty — ascribed to himself. His theory, psychoanalysis, clarified that even in his last fortress, his own mind, man did not reign nor dominate. Man does not necessarily know the depths of his soul better than others do, or better than he knows the external world.

Freud first spoke of Copernican revolutions in a series of lectures he gave in 1914 in the University of Vienna. At that time neither he nor, for that matter, anybody else, was aware of a fourth Copernican revolution in the making, more significant than all its three predecessors put together. This revolution deprived man of his exclusiveness in the kingdom in which he was considered not only as a sovereign, but also the sole inhabitant: abstract thought. From the middle of the 19th century the understanding started permeating that abstract thought, like the motion of bodies or flow of liquids, is a physical process that can be described in mechanical terms and may be performed by machines.

Not surprisingly, the seeds of this revolution were sown in England during the industrial revolution. The victory of the machines induced thinkers to ascribe mechanical behavior to all phenomena, including human ones. Some, like Karl Marx, who applied deterministic concepts to social processes, were theoreticians. Others tried to put it to practice. In 1820 Charles Babbage designed a calculating machine, whose great advantage was that it could be programmed. Although the project got huge financial support from the government it was never completed. The actual materialization took place only one hundred and fifty years later, in the London Science Museum, and the result was proof of Babbage's genius. The machine actually worked.

The greatest victory of the fourth Copernican revolution was in mathematics. Mathematical thought is the most organized form of human thinking, and therefore it is relatively easy to describe in mechanical terms. The mathematical branch that does it is called "mathematical logic". A modern pioneer was the Englishman George Boole (1815–1864), who studied the ways statements join together by connectives, and the rules that govern these connectives. For example, statements can be joined by "and", which Boole denoted by "$\wedge$": $(3 > 2) \wedge (3 + 4 = 7)$ means that $3 > 2$ and $3 + 4 = 7$, which is true since both components are true. "Or" is denoted by "$\vee$", and so $(2 > 1) \vee (1 + 1 = 3)$ means "$2 > 1$ or $1 + 1 = 3$", which is again true, since for "A or B" to be true it suffices that one of them is true. Negation is denoted by "$\neg$". For example "it is not true that $2 = 1$" is written "$\neg(2 = 1)$". Implication is denoted by an arrow: $(2 > 1) \rightarrow (1 + 1 = 3)$ means "if 2 is larger than 1 then $1 + 1 = 3$" — by the way, an obvious fallacy, since 2 is indeed larger than 1, and yet $1 + 1$ is not 3.

This way of viewing mathematical thought demands a shift of position: from that of a participant in the game of thinking to that of a detached observer. Uttering statements and reacting to them is replaced by relating to them as objects in the world, strings of symbols on paper. Nowadays this seems self-evident, since this is the way we look at computers. In Boole's time it was a conceptual leap.

The Cement of Mathematical Thought

Statements are the building blocks of mathematics, but blocks do not constitute a building. Something must cement them together into meaningful structures. This role is performed by *proofs*. Proofs are the pinnacle of mathematical activity. They ascertain what is true and what is not, but more importantly — they are testimony to understanding. "Proofs are not there in order to show that a theorem is true. They are there to tell us why it is true," said the American mathematician Andrew Gleason.

It was a German philosopher-mathematician, Gottlob Frege (1848–1925), who took the next daring step: showing that proofs, too, can be formally studied. The first thing he did was add another connective to Boole's notation. It stands for "for all", and after some vicissitudes its notation has stabilized to be "$\forall$", alluding to "all". For example, $\forall x(x > 7)$ means "every number is larger than 7" (a manifestly false statement, when interpreted in the natural numbers). In 1897 an Italian named Giuseppe Peano (1858–1932) introduced another notation, for "exists". Again, the notation underwent many changes, and today, quite naturally, it is denoted "$\exists$". For example, $\exists x(x > 7)$ means "there exists a number larger than 7" (an obviously correct statement, when interpreted in the natural numbers. For example, 8 is such a number).

Using the new notation, it was possible to express any known mathematical statement, and Frege could now embark on his next mission: defining "proof". Everybody who has ever seen mathematical proofs is probably familiar with their foremost property: linearity. They are constructed brick upon brick. Indeed, this was Frege's definition of "proof":

A proof is a sequence of statements, each following from its predecessors.

By itself, this is no novelty. The layered character of proofs is, as mentioned above, well known. The surprising insight of Frege was that "following" is a simple notion. The game of proof is not as complex as it may look. It is governed by only two deduction rules.

The first of these was formulated already by Aristotle, and is called "*Modus Ponens*": If you know that both P and $P \to Q$ (here P and Q are any statement) are true, then you can deduce Q. If you know for sure that today is Tuesday, and you know also that "if today is Tuesday then it is raining", you can deduce that it is raining. Formally:

If P appears somewhere in the list of statements constituting the proof, and $P \to Q$ appears somewhere else in the list, then you can add to the list, as the last current line, the statement Q.

The second rule is also very simple, but is a bit surprising. It is called "Generalization", and it says the following. Suppose that you have proved that John is mortal, and suppose that you did not use any special property of John's apart from his being a human. Then you can deduce that every human is mortal. After all, the proof is valid for any human being. Formally:

If a statement P appears somewhere in the proof, you can add the quantifier $\forall$ ("for all"), namely write $\forall xP$.

In fact, for our purposes it is not important what the deduction rules are. It only matters that they are simple and few. It means that the game of proof is simple. It is easy to recognize a proof when you see one. Of course, concocting proofs is another story, and may be much harder, just as it is easier to recognize a familiar person when you meet him than to find him in a big city.

Each of the two deduction rules proves statements from previous ones. But clearly, this is not enough. There must also be a starting point. The first line in the proof doesn't have any preceding lines to fall back on. Famously, earth stands on a giant tortoise, but what does the tortoise stand on? We have to start somewhere, from rock solid statements that are true in their own right. These are the **axioms**, basic facts that are not open to doubt. Frege formulated some basic axioms, for example $P \to P$ (if today is Tuesday then today is Tuesday), or $\neg\neg P \to P$ (if it is not true that today is not Tuesday then today is Tuesday). On top of these general axioms, each field has its own. For example, in geometry one of the axioms will be "one and only one line can be drawn through two distinct points" — a statement that should of course be translated into formulas. In number theory, it is natural to add as an axiom "adding 0 to a number doesn't change it", or formally: $\forall x\ (x + 0) = x$.

So, Frege's definition of a proof is this:

A proof is a list of statements, each of which being either an axiom, or a derivation from two of its predecessors by *Modus Ponens*, or a derivation from one of its predecessors by Generalization.

To complete the definition one more explanation is needed: what does the proof prove? This is easy: its last statement, the punch line.

Meanwhile, Across the Channel ...

I have told you about the opposition Cantor had to face. Frege's discoveries met a fate worse by far: almost total disregard. One of the very few who noticed them was Cantor, who wrote in 1880 a vituperative review. Twenty years elapsed before he won recognition, and it could have taken longer, but for the lucky fact that Bertrand Russell had a German governess when a child. By the age of four Russell had already lost both his parents and his elder sister. He and the other remaining brother, Frank, were brought up by their strict grandparents, who hired a German governess to educate them. This is how young Russell came to know German well enough to become familiar with the writings of Frege around 1900. He immediately recognized their importance.

When Bertrand was 11, his brother Frank introduced him to Euclid's geometry, and this changed his life. Not only because of the love for mathematics that he contracted, but also because of the interest it awoke in him in the notion of "axiom". "Why should I believe the axioms?" he asked Frank, and instead of an answer he received an admonition — "If you don't accept them then we just can't continue," causing young Russell to take an oath to understand what makes a statement true.

In the late 1890s Russell initiated a project similar to that carried out by Frege over a decade earlier. Like Frege, he started writing a book on the foundations of number theory, and reached similar conclusions. Having then learnt about Frege's work, he joined forces with his teacher in Cambridge, Alfred North Whitehead, and together they embarked on a venture of colossal dimensions: writing parts of the mathematics of their time in Frege's new language. They showed that mathematics can be written formally, instead of in spoken language. The result was the three-volume *Principia Mathematica* published between 1910 and 1913. The name is presumptuous: this was the name Newton gave his monumental book from 1687 (the full name of the latter is *The Mathematical Principles of Natural Philosophy*, namely of physics). The *Principia* is one of the best-known books in the history of mathematics, and justly so, and also one of the least read books, which also stands to reason. There is no real point in reading the details of the grandiose enterprise. Suffices it to know that it can be done.

A great gift was bestowed upon mathematics — a formal tool. In principle, it was now possible to take any mathematical argument and express it in formulas instead of words. The translation lengthens the argument many times over. A few lines of verbal argument turn into pages and pages of barely digestible formulas. Nobody intends to do it for real — mathematical papers are still mostly words. But it is important to know that the possibility exists, if only for the fact that written formulas can be handled by the computer. At the present state of artificial intelligence, computers still do not understand natural language well enough to be able to follow mathematical arguments, but they can understand formulas. And they are not daunted by length.

What should be done with the tool with which the wisdom of Frege, Russell and Whitehead endowed the world? On this point there wasn't much hesitation. Frege already pointed the way in his book *The Foundations of Arithmetic.* First, he said, the tools of logic should be applied to number theory — after all, numbers are the very heart of mathematics. A simple system of axioms for numbers should be formulated. And indeed, such a system existed at the time. In 1888 Peano suggested a set of nine natural axioms that is customarily referred to as PA (Peano Axioms). For a long time PA was considered exhaustive and comprehensive. The assumption was that it is the last word, capable of proving every correct fact about numbers. But nobody knew how to prove that this assumption is correct.

A Very Ambitious and Very Wrong Program

Hilbert was one of the last mathematicians who knew in depth a large part of the mathematics of his time. He had a special mode of work. He would go into a field, plough it for a few years, make fundamental contributions, and move on to the next field. His ties with women followed the same pattern, only with more frequent transitions. Just as he was one of the first to understand the significance of Cantor's set theory, he understood before many others the potential of the developments in logic. Trying to point the way for his fellow logicians, he posed four fundamental problems, four tasks that jointly gained the name Hilbert's Program.

1. **Completeness**. The first task was proving that PA, the set of Peano Axioms, is complete. Namely, that using it one can prove all true statements about numbers. If you conjecture, for example, that every even number is the sum of two prime numbers (the famous Goldbach conjecture), PA should either prove or disprove its truth. And so it should be able to do for any mathematical statement.
2. **The decision problem**. Can a computer program distinguish between true and false facts about numbers? If we assemble the best mathematicians in the world and the best programmers, will they be able to write a program that decides, for any given statement, whether it is true or not? Of course, in Hilbert's time there were no computers, and instead he spoke of an "algorithm" — a recipe for a decision-making procedure. At the time there was

also no precise definition for "algorithm", but Hilbert did not consider such a definition to be necessary. When you have an algorithm at hand you recognize it as such. And since Hilbert believed that a decision algorithm will be found, he did not bother about the definition.

3. The third task was similar, but not identical: **deciding on provability**. Is there a computer program that, given a statement, can decide whether or not the statement is provable in PA? Since "correctness" and "provability" are not necessarily the same (the first task asks if they are — but, as we shall soon see, they are not), this is different from the previous task.
4. The fourth question was the most delicate. It relates to the possibility of **contradiction arising from PA**. Is it possible that two contradictory statements, like "10 is an even number" and "10 is an odd number" could both be proved from PA? Actually, we know that this is impossible: Peano chose his axioms carefully, so that whatever is provable from them is correct. So, if this were the case, 10 would be both even and odd, which is impossible. But Hilbert asked something much more subtle: can you prove that there is no contradiction, just by contemplating Frege's deduction rules and the axioms of Peano? Looking at the rules of Tic-Tac-Toe you can deduce that each of the players can force a draw — can you similarly study the rules of the game of proof and guarantee that they do not lead to a contradiction?

Hopes soared. A new era in logic was anticipated. And indeed, such an era came. Only it took a diametrically opposite direction. Hilbert presented his program in Bologna in 1928, and Gödel was attending. At the time he was planning to study physics, but the lecture changed his life. For two years he sat locked up in his room, struggling with Hilbert's problems, and when he eventually emerged it was with a sensation — a refutation of the program, in all its details. All four tasks turned out to be impossible.

1. **The Incompleteness Theorem**: PA is not complete. There are statements about numbers that are true in the natural numbers, but cannot be proved using only the Peano Axioms.

2. **Undecidability (the demise of the decision problem)**: There is no algorithm (= computer program) that for any given statement on numbers can decide correctly whether it is true or not.
3. **No decision algorithm for proofs**: There is no algorithm that given a statement on numbers decides whether it is provable from PA.
4. **No proof of consistency**: There is no "syntactic" proof (namely a proof reached by analyzing the rules of proof and the axioms of PA) that PA is consistent. That is, it is impossible to prove using only syntactical tools that PA cannot prove a statement and its negation. The game of proof is too complicated to analyze in full. Number theory is more complex than Tic-Tac-Toe, or even checkers or chess.

These are four different results, each forming a world of its own, and all this was proved in about 30 pages. The world needed some time to digest the new theory, but once it did the implications were far reaching.

Gödel's Paradox

The proof of Gödel's theorems — for indeed it was basically one proof behind all of them — was born of a paradox. In fact, as Gödel himself testified, two paradoxes: the Liar, and a paradox parodying Cantor's Mr. Spite proof, called Richard's Paradox. The first gave Gödel the basic idea, while the second gave him an idea that we shall soon meet, of numbering formulas.

But Gödel didn't simply use existing paradoxes. He formulated a new one, that is more complex than all logical paradoxes that we have met so far. Look at the following statement, call it G.

G: This sentence is not provable.

Assume for a moment that G is false. Then, by its content, it is provable. But, as everybody knows, if something can be proved, then it is true. After Pythagoras proved his theorem, there is no point in trying to find a counterexample, namely a right-angled triangle that does not satisfy it.

What have we shown so far? That if G is false then it is true. But this means that G cannot be false. Its falsity would entail a contradiction — something (itself) is both true and false.

So, G must be true.

But look at the last lines: they constitute a proof of G! We have just proved that G is true. So, it is provable.

So, by its content, G is false. After all, it says that it is not provable, so its provability means that it is false.

We have thus shown that G is both true and false, a paradox.

On the face of it, Gödel's paradox is similar to the Liar Paradox. But the resemblance is only superficial. The deceptions in the two are very different. The deception in Gödel's paradox is much subtler.

So, where is the cheating? When you want to crack a fraud, a good idea is to go to the part of the argument that is declared as "self-evident". Here it is the declaration that "whatever is provable is true". It is not as innocent as it looks. It cannot be used as a rule in proofs. "Every provable statement is true" sounds natural and self-evident, but it **depends on the rules of inference**. If these rules are inane, then this is false. For example, if one of the inference rules is that any sentence starting with the letter "A" is true, or that any sentence uttered by the prime minister is true, you will prove many absurd statements. This means that the validity of the rule that what is provable is true is very much dependent on the choice of the rules. But then, if it itself serves as one of the rules, as happens in the paradox, then what we have just said is that "the rule is valid, under the condition that it is valid". Its validity is circularly defined.

How can we know that this is indeed the tricky point? Here we can be aided by the mathematical version. Gödel didn't formulate his argument as a paradox, but as a mathematical proof. He wrote a number theoretic formula that says about itself precisely this: "I am not provable". In the mathematical version there is no contradiction. So, to be wiser we should follow the argument in the paradox, and find **which of its steps does not go over to the mathematical proof**. I will do it in *Part VII For the Experienced Hikers*, and I will show there that the stage where the argument in Gödel's paradox breaks down in the mathematical version is precisely this point. There is no logical axiom saying "if something is provable then it is true", and there cannot be such an axiom.

From Paradox to Proof

To go from paradox to theorem, Gödel had to perform a daredevil leap. Those who knew him in person probably wouldn't describe him as "bold" (unless ignorance of social codes counts as boldness). True, he went through some non-trivial adventures, like fleeing Austria through Japan and Russia to get to the United States, after some hooligans attacked him in the street, because he "looked Jewish". But this he did not do by choice. The boldest thing he did of his own volition was marrying a cabaret dancer six years his senior, in opposition to his mother. But life and science are not mirror images of one another. The idea of how to go from the paradox to the proof of the Incompleteness Theorem is more than bold — it is insanely brave. You need courage to believe that a sentence like "this sentence is unprovable" can be written as a formula about numbers. This is precisely what Gödel did. He designed a formula in number theory that says about itself that it is unprovable.

You may well find this hard to follow. A formula in number theory cannot "say" anything about itself, or about provability. It can only say things about numbers. The key idea of Gödel that made it possible for a formula to speak about itself was assigning numbers to formulas. Put more simply — numbering the formulas. With every formula he associated a distinct number. There are many ways of doing this, the simplest being just counting: formula number 1, formula 2, formula 3, and so on. For example, you can order the formulas by length, from short to long. For a given length, order the formulas of this order in lexicographical (dictionary) order, assuming some alphabet-like order on the logical symbols. But this is very cumbersome. One problem is that given a number does not make it simple to recognize the formula: which formula will be numbered a million and 1? Gödel chose his numbering much more cleverly, in a way that given a number it is easy to reconstruct from it the formula assigned to it (if one is assigned — not every number is used).

Having numbered all formulas, Gödel could now achieve his goal. He could write formulas that speak about formulas. They do it circuitously: when you want to say something about a formula, just say something about its number. And indeed, Gödel showed that using his numbering a property like "a formula is provable" can be nicely translated to a property of the number of the formula. You can write a formula about a number n, that is true if and only if the formula corresponding to n is provable.

In other words, there is a formula, call it $\pi(x)$ (π, the Greek "p", for "provability"), that is true if and only if x is the number of a formula that is provable from the axioms of Peano.

Now comes the crucial step: constructing a number theoretic formula G, whose number is, say, x, that says "not $\pi(x)$". This means that G has the following property:

G is true if and only if it is not provable from the axioms of PA.

I am not saying that constructing G is easy. In fact, it seems that the construction is a circular task, since you first need to know its number x, and only then formulate G. But with some technical ingenuity, this is doable. And from the moment that we have G at hand, the way to the proof of the Incompleteness Theorem is short. It follows almost verbatim the argument of the paradox. But this time we reach a theorem, not a contradiction. Here is the argument.

Assume that G is false. By the property of G, this means that it is provable from Peano Axioms. But as we mentioned in the chapter A Very Ambitious and Very Wrong Program, Peano's Axioms are valid, meaning that anything that can be proved from them is true.

This shows that if G is false then it is true.

But this implies that G **must be true**, or else a contradiction has arisen.

But by the above property of G, its being true means that **it is not provable**.

So, we (or rather Gödel) produced what he set out to find: a formula that is true but not provable. We have thus proved the Incompleteness Theorem:

> **There exists a number theoretic formula (viz. G) that is true but is not provable from Peano Axioms.**

Why didn't we reach here a contradiction, as in Gödel's paradox? Because the punch line of the paradox is missing: "The last lines form a proof of G, showing that G is provable, contrary to what we have just shown." But why don't they? Haven't we just proved G, black on white? Indeed, we have. The reason is that the proof was **not formal**. It was in English. And we have only shown that G is not provable formally, from the axioms. It may well be, and actually is provable in English.

Surprise Tests

As I told you, new paradoxes are rare. For this reason, when a seemingly new paradox was published around 1950, it was an event worth notice, especially since it was a very attractive paradox, taken almost straight from real life. It was born in Sweden, during the Second World War. An announcement on the radio went, "Next week there will be an unexpected air raid alarm drill." A mathematician named Lennart Ekbom noticed the paradoxical implication of this announcement. Nowadays the paradox is known as The Surprise Test or, dressed in a different story, the Hangman Paradox. But, in fact, the paradox was not really new. It was just Gödel's paradox in new attire.

> A teacher announces to his students that next week there will be a surprise test. He also defines what he meant by "surprise". It means that the evening before, the students will not expect it. The students now argue: the exam cannot take place on Friday, since in this case on Thursday evening they would know that it will be tomorrow, Friday being the only remaining day. Friday being marked off, the last possible day is Thursday. So, if the exam does not take place until Wednesday evening, the students will know that the exam is tomorrow. So, Thursday is also out. Continuing this way, the students deduce that the exam cannot take place on any day. But then, on Tuesday, the teacher tells them to take out pen and paper, and they are totally surprised.

Of course, the story smells of circularity. The argument is "the teacher knows that we know that the teacher knows ..." — eventually the knowledge must refer to itself, but how, precisely? One rule in solving problems is — go to extremes. Consider the most extreme case that you can think of. Martin Gardner, the science popularizer whom we already met in the chapter on Newcomb's paradox, suggested considering the case of one day, instead of six. In this case the teacher says just this, "Tomorrow there will be a surprise test." Gardner also replaces the story by:

> A husband says to his wife, "Tomorrow I will surprise you with a beautiful necklace." She now argues: he cannot fulfill his promise, because part of it is the surprise. So, he will not buy me the necklace. But now I know he won't, so if he does it will be a surprise. So, he can fulfill his promise. But now it is not a surprise, because I know it — and so on and so forth.

The paradox, says Gardner, is nothing but:

S: You don't know this sentence.

The exam and the necklace are nothing but distractions. We already met this idea in the story about Smullyan's brother, who promised him on April Fool's day to trick him, and didn't (or was that the trick?). But this means that this is precisely Gödel's paradox, with "unprovable" replaced by "not known". "Prove" and "know" are interchangeable in this context, since "prove" means "make sure you know", and the "know" is "prove to yourself". The sequence of arguments is just as in Gödel's paradox:

> If S is false, then (since it says precisely that I don't know it) I know it. But everything I know for sure is true. Thus, if S is false then it is true. So, it must be true. Having proved S, we now know it. So, by its content, it is false.

And so on and so forth — we have reached our endless circle. And if the paradox is identical to Gödel's paradox, so are their solutions. The deception is in "if we know something then it is true". This cannot serve as an axiom in logical deductions.

The Theorem of the Century?

When Gödel received an honorary Ph.D. from Princeton University, the document describing his achievement dubbed his theorem "the most important theorem of the 20th century". But as formulated above, it hardly deserves the title "theorem of the month". "Peano Axioms are not complete" — so what? So, one particular system is incomplete — not a big deal. All it probably means is that Peano was negligent. Somebody else can come along and devise a better system that **will** be complete. If we only add enough axioms, we shall probably obtain a complete system, proving every true statement. For example, we know that Gödel's formula G is true, why not add it to the axioms? Perhaps this will result in a complete system.

It will not. Not if you add one axiom and not if you add a million. After we add G, there will be another true formula that is not provable, and when we add this one there will be another. For, Gödel proved much more than just that PA is incomplete. His theorem is much further reaching:

No reasonable axiom system for numbers is complete.

What does "reasonable" mean? It means satisfying a requirement that is indeed very reasonable to pose: that we shall know what is an axiom and what is not. Clearly, a system of axioms is useless if we do not know how to distinguish axioms from non-axioms.

But what does it mean "to know what is an axiom"? It means, according to Gödel, that there is a computer program — an "algorithm" in Gödel's language — that recognizes axioms. Given a formula, this algorithm should be able to decide whether the formula is an axiom or not. For example, PA is such a system. It is "reasonable". In fact, any system that can be written explicitly is reasonable. Once we have the list in front of our eyes, it is easy to write a program that recognizes its members.

So, Gödel's theorem speaks about every axiom system worth considering. If your system is simple, in the sense that you know what is an axiom and what is not, it will not prove all true facts about numbers. For simplicity you pay with incompleteness. The truth about numbers cannot be captured by explicitly written axioms — this is indeed a very general theorem, deserving the title the theorem of the century.

Here a moot point arises, that will connect us to the second part of Gödel's theory — the decision problem. Every reasonable axiom system is incomplete, we said. But is it really so? Here is an axiom system that is perfectly complete, and seems to be completely perfect. Let us call it T, and it is defined as follows:

T is the set of all true formulas.

"True" is in the natural numbers. For example $1 + 1 = 2$ is in T, and so is $\forall x \exists y \ (y > x)$ ("for every number there is a larger one" — an obviously true statement). T is obviously complete, namely it can prove every true statement. Given a true statement, it is in T, and T just has to declare "it is an axiom", and thus present a (one line) proof.

On the other hand, T is so easily definable — what can be simpler than "a formula is in T if it is true"? And doesn't this mean that T is "reasonable"? If so, this contradicts Gödel's theorem that reasonable axiom systems are incomplete.

Well, don't worry. It is impossible to give a counterexample to a theorem rigorously proved, and Gödel's theorem was proved rigorously. All it means is that T is not "reasonable". Why? We have just proved it. It is complete, and by Gödel's theorem a complete axiom system cannot be reasonable.

Now, let us look at what this means. The non-reasonability of T means that there is no algorithm that tests which formula is in T and which is not. Since T is defined as the set of all true formulas, this means that:

> **There is no algorithm that checks for every formula whether it is true or not.**

But this is precisely part 2 of Gödel's theorem, as mentioned in the chapter A Very Ambitious and Very Wrong Program. It says that number theory is "undecidable": it is too complex to be grasped by a single computer program. No single computer program can decide the validity of every possible formula.

Keep in mind that this relates only to number theory. Other theories can be decidable, namely Gödel's theorem is not necessarily true for them. For example, the Polish mathematician Alfred Tarski (1902–1983) proved that plane geometry is decidable. This means that there is a computer program, which given any statement in geometry outputs "true" if the statement is true, and "false" if it is not. Geometry is not as complex as number theory.

Why it was Imperative to Mummify Lenin

When philosophical ideas associated with science are dragged into another field, they are usually completely distorted.

(Richard Feynman, *The Feynman Lectures on Physics*)

Philosophers seized on Gödel's results with great fervor, usually with total lack of understanding. Probably no other scientific idea has bred so many inanities. In the book *Intellectual Impostures* by Alan Sokal and Jean Bricmont, on the scientific pretensions of the French post-modernists, there are some amusing examples for "applications" of the theorem. Proofs, using the Incompleteness Theorem, of the inability of a nation to rule itself; the impossibility of a society to be free of mysticism; an explanation why it was compulsory to mummify Lenin and support for Spinoza's proof of the existence of God. Wittgenstein outdid them all. He looked at the giraffe of Gödel's theorem and said "there ain't no such animal". He claimed that the theorem is false, on the grounds that it contradicted some of his own views. Modesty was not one of his failings. At the end of his thesis defense he clapped Russell and Moore, the two examiners who procured him a position in Cambridge, and told them, "Don't worry, I know you will never understand it."

And yet, there is a recurring philosophical argument that seems to deserve consideration. It is "the advantage of man over the machine". By Gödel's theorem, says this argument, machines have limited power. There are things that a machine cannot perform. Man, by contrast, does not have this limitation.

Indeed, Gödel's theory is all about impossibility. "There are no algorithms for doing this and that." The original impossibility theorems were about deciding on formulas, whether they are true or not, or provable or not. But from these there emerged a whole body of impossibility results. Here is an attractive one: suppose you are given a finite number of planar shapes. Can you tile the entire plane, using as many copies (possibly infinitely many) of each shape? No machine can decide whether this task is possible for every set of shapes.

This limitation, says the argument, does not apply to man. Man is creative. Whereas the computer is limited to what its program tells it to do, man can come up with new ideas, he can innovate. He will tackle each problem (say, each formula he meets) with new tools.

But this argument does not carry much weight. It is like realizing that a racing car cannot exceed a certain speed, being only a machine, and concluding that man has the advantage that his body is "creative", and can overcome any technical limitation. Man's brain is a wondrous machine, but just like the body it has its limitations. It is a machine that operates by certain rules, and is as limited as any other machine.

Part VI: Turing Invents the Computer

The rabbit felt mighty important that day
On top of the hill in the sun where he lay.
[...]
On land, and on sea ... even up in the sky
No animal lives who is better than I!

(*The Big Brag*, Dr. Seuss)

From Olympus Down to Earth

News about the revolution in Vienna spread quickly. And history now repeated itself — the next step was again taken across the channel. In 1934 a mathematician named Max Newman gave a course in Cambridge on Gödel's discoveries. Among his listeners was a young student by the name of Alan Turing (1912–1954). Having mechanical inclinations, the bright young mathematician was particularly attracted to that side of the theory that dealt with algorithms. He decided to give them concrete meaning, namely to let them be performed by real machines. Gödel's abstract definition of "algorithm", so he was determined, should be replaced by tangible cogwheels. He succeeded in doing this, and thereby supplied the missing link between Gödel's ideas and the modern computer.

Turing came up with a model of a machine that performed calculations, namely it receives numerical input and produces numerical output. Both input and output are written in the binary system, namely with 0, 1 digits. The model is surprisingly simple. Its main ingredient is a tape, divided into cells on which the input given to the machine is written, one symbol in each cell. The machine uses this tape also for writing the output.

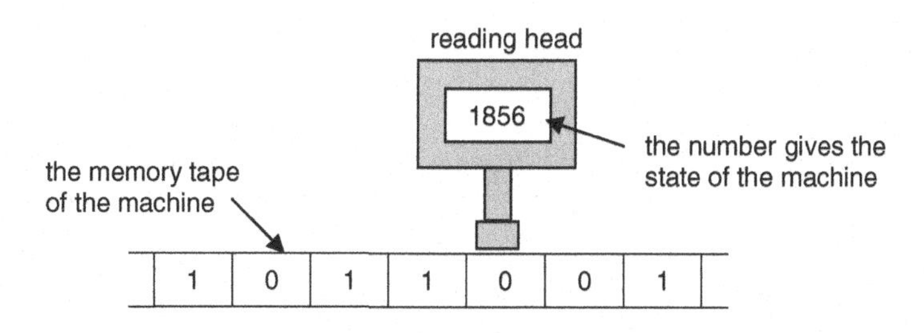

Like the modern computer, Turing's machine is programmable, namely it has a set of predetermined instructions. At each step the machine is in one of a finite number of "states": just as a modern day computer is at each point in time in some physical state, determined by what is stored in its memory. And the program, with which the machine is endowed, tells it to what state it should go from the current state. So, a typical line in the program is: "If you are at state number 5775 and in the cell at which you are now the digit written is 1, then replace the 1 by a 0, move one cell to the right, and switch to state number 4321".

The machine is given an initial input. There is a special state, called FIN, that if the machine enters it, it has to stop. Whatever is written on the tape at this point is the output of the machine.

So, what does a Turing machine do? Outputting a number for any input number means an assignment of a number to every number. Such an assignment is called in mathematics a "function", and so a Turing machine calculates a function. Turing proved something surprising: that the functions that his machines can calculate are precisely those abstractly defined by Gödel's algorithms. And then something even more intriguing happened: in the very same year that Turing published his theory, 1936, two other mathematicians, Alonzo Church and Emil L. Post, invented two other models of calculating devices. And lo and behold — both turned out to be of the same power. They also could calculate precisely the same functions. This persuaded most mathematicians that an "algorithm" is precisely what can be performed by a Turing machine. Eighty years have passed, and this has not been refuted. Nobody has been able to come up with a more powerful mechanism of calculation.

The Halting Problem

Gödel proved that there are problems too complex to be solved algorithmically. But these were abstract problems, like deciding whether a statement on numbers is true or not. Turing searched for a "real life" problem that is not programmable. And he indeed found one, a problem that does not deal with numbers, but with machines — his own machines. It is called "the halting problem".

Consider the (almost) simplest of all Turing machines, call it IDLE. It has only two possible states, the beginning state START and the terminating state FIN. The only command it has is: "Whatever you read in the cell you are at, don't change your state, don't move and don't change the symbol in the cell." If not the simplest machine, it certainly is the laziest one. What will happen? Nothing of course. You can leave the machine on your desk, go do your business, and it will run forever. It will never stop. Compare this with the (really) simplest machine of all, one that has just one command: "Go to state FIN, not changing anything." This machine will halt after one step. So, there are machines that stop, and machines that never do.

In the examples above it was easy to decide which of the two possibilities occurs. Had the descriptions of the machines been more complex, it would not have been so easy. This is the "halting problem".

> **The Halting Problem**: Given the description of a Turing machine and an input with which it starts, decide whether the machine, with this input, halts or not.

Turing proved that this decision task is indeed not easy at all. In fact, there is no algorithm that can perform it. In modern day language: there is no single computer program that, given as input a computer program P and its input I, decides whether P stops if it starts with I.

Of course, Turing did not speak of computers. In less than ten years he would be going to invent one, but we are still in 1936. Instead, he spoke about his machines. He showed how the description of one machine (say S) and its tape input I can be given to another machine (say T). With this information given to T, we can ask T: "Does S halt, when starting with input I?" Turing proved that such a machine T cannot be constructed. Deciding whether a given machine halts on a given input is too complex a problem to be determined mechanically.

How did Turing prove this? We should not be surprised: through circularity. He showed that there are cases in which the task of decision becomes circular, and hence impossible. Supposing that a machine T as above exists, we could let it have as input its own description. And then, if it halts on this input, it is easy to go on and tell it — go roaming, like the machine IDLE described above. And if it decides that it is not halting, tell it to stop right away. In short — tell it to do the opposite of its own prediction. Since the input was T itself, any prediction of itself would turn in this way into its own refutation. This is not far from the Liar Paradox.

The undecidability proofs of Gödel and Turing were the glorious victory of the concept of self-reference. They closed the brackets opened in 1878 by Cantor, with his Mr. Spite proof. In the fifty some years in between mathematicians understood much better than ever before the nature of their occupation. Mathematical logic provided the understanding of what is mathematics; circularity clarified its limitations. So, one of human mind's greatest victories was the understanding of its own limits.

The Imitation Game

The decade following the publication of his groundbreaking work was for Turing a period of hectic activity. He spent two years in Princeton, where he wrote a thesis on non-computable functions. At that time he believed that these functions would shed light on human intuition, which is also "not programmable", so he believed at that time (he would later change his mind). Then he returned to England, to what would be the best-known chapter in his career — Bletchley Park, the center for code breaking. He was a key figure in breaking the code of the Enigma machines, which served the Germans to communicate with their submarines. For the massive calculations involved, Turing and his team constructed the Colossus, the first calculating machine that deserves the name "computer". Before, Polish mathematicians tried to break the Enigma code using primitive computing machines called "Bombas", but the Colossus was much closer to modern computers.

After the war Turing's life lost its focus. He was involved in some projects of developing computers, but his teams lost the race to the Americans. Because of his homosexuality he was excluded from top secret security projects. He then turned to biology, in which he wrote some seminal papers that are studied until this very day. In parallel, he tried to bring to the awareness of the general public the new Copernican revolution, incarnated in the invention of the computer. For this purpose he concocted a thought experiment that came to be known as Turing's Test. He published this experiment in 1950 in a paper entitled "Computing Machines and Intelligence" that appeared in *Mind*, the foremost philosophical journal.

Suppose, says Turing, that in one room there sits a man, and in another a woman. You can ask them in writing any question, without seeing or hearing them, and by their answers you want to find out which one is the man and which is the woman. The woman wants you to reach the right conclusion, but the man wants to fool you. Can he achieve his aim? Obviously, he can. Now, asks Turing, what happens if instead of a man-woman pair you have a machine-human being pair? Can you decide with certainty which is the computer and which is the human being, just by asking them questions from outside the rooms?

The first point that Turing was making was that the only meaning of "can machines think?" is in the outcome test. It is not the composition of the machine that matters, its hardware or its origin. Thinking should be judged by its output. In this he echoes some Wittgensteinian ideas, who claimed that "the meaning of a word is its use".

Turing goes on to refute some of the most common objections to this idea. He lists the things that people claim computers can't do:

> ... be generous, beautiful, friendly, resourceful, have a sense of humor, tell right from wrong, make mistakes, fall in love, enjoy strawberries and cream, make someone fall in love with it, learn from experience, use words properly, have as much diversity of behavior as man, be the object of its own thought, do something really new.

"No support is usually offered for these statements," claims Turing. And now, sixty years later, more can be said — there is by now plenty of support to the contrary. For some decades contests have been held among computer programmers, of succeeding to fool humans in the Turing test. In 2014 a computer program managed to convince a panel that it is a Ukrainian boy of 13. It is nowadays easy to program a machine so that it will have feelings. For example you can program a computer that writes "I love you" to his programmer, blushes when the programmer arrives (you should provide the computer with blood circulation for that, but this is a technical matter), make deliberate mistakes so that the programmer stays overtime. "It is not love, it is just mechanical reactions," you may object. But Turing's message is that the reactions are the love.

Will the computer ever be able to perform the most abstract tasks man can do, like writing good poetry, or good music, or making scientific discoveries? I myself do not doubt that this will happen one day. When computers will think billions of times faster than humans, there will be no need for human scientists. Should we be glad or sad? This is not the question, for it will happen. And when it does, it will be thanks to the founders of mathematical logic that showed us how to describe thinking in concrete and formal terms.

Searle's Chinese Room

Four years after the publication of "Computing Machines and Intelligence" Turing committed suicide, after having been convicted of homosexuality and punished by chemical castration. This was not the first time that puritan England killed one of its most ingenious sons. The dialogue of Turing with the philosophical world was cut short, but one-sided debate did continue. In 1980 an American philosopher, John Searle, presented a thought experiment of his own, a kind of parody on Turing's imitation game. In Searle's story, too, somebody is sitting in a sealed room and is trying to convince the outside world of something he is not. But Searle adopted a point of view different from that of Turing: he was looking from the inside.

Assume, said Searle, that you are sitting in a closed room, and an experimentalist is showing you through a window signs of the Chinese language. You have no knowledge of Chinese, and you do not have a clue what these signs say. But in the room, hidden from the eyes of the experimentalist, you have a book with instructions, telling you that if you see such and such a sign, you should raise on your side a board with this and that sign. Perhaps if the experimentalist raises a board with a sign saying "dog" you are supposed to raise a sign saying "cat" — when, of course, you do not recognize either sign. You yourself know that you do not understand a thing, but for the outside observer you appear to know Chinese.

The moral is clear. The computer is the entity that reads signs, pretending that it understands, feels, thinks, or speaks Chinese, when in fact it does not understand anything. It is nothing but a lump of metal with electrons running in it to and fro. It does not really think, but only moves cogwheels in its "brain". Its motions are nothing but empty gestures.

How would Turing have answered Searle had he been alive? One can learn that from his answers to the questions on the computer's inability to feel and to sense. He writes that the idea that man thinks in a different way is based on an assumption that he knows his thinking in a different way from that in which he knows the thinking of the computer, namely that he has private access to his inner world. This, says Turing, leads eventually to solipsism, namely the belief that only he exists, and the rest of the world is but a projection of some sensations on his screen of consciousness. And such a belief, says Turing, is a heavy price to pay.

I can guess that Turing would answer Searle that, looking from aside, there is no difference of principle between a native Chinese speaker and the owner of the instructions book. The Chinese speaker, too, has a book of instructions that tells him how to react to certain phrases and situations. The difference is only in the amount of internalization. The instructions in the brain of the native Chinese speaker are so well internalized, that he is not aware of them. But this is only a quantitative, not a qualitative, difference. The reader of instructions, on the other hand, is aware of the mechanism that governs his actions. It is precisely because of the lack of internalization that he is wiser — not in Chinese, but regarding the question on the difference between man and machine.

Of course, for an interpreter I would choose the native Chinese speaker, just as I would elect to go to a restaurant run by an experienced chef rather than to one run by children taking their recipes from the book *Children Cooking*. But when looking for somebody who understands better what it means to know Chinese, I would choose the instructions reader.

Personality and Genius

The film about Turing, *The Imitation Game*, suggests that he was on the border of autism. *Imitation Game* refers not only to Turing's test, but it also suggests the possibility that Turing behaved mechanically, imitating the way "real" people, motivated by feelings and emotions, act. Gödel was not a simple person, either. He was very quiet, scared of confrontations, and his human relationships were meager and bizarre. He was totally dependent on his wife Adele, a somewhat common woman. When she fell ill and was hospitalized, he did not trust anybody else's cooking and starved himself to death.

Does genius come to the world bound together with strangeness? This claim can be heard over and over again. Biographies of geniuses often stress their pathologies. Is there really a connection between the two? I tend to believe that there is no real link. The gift of genius can be given to balanced people just as much as to unbalanced ones. More than that: scientific achievements usually require a stable personality. But in the case of Gödel and Turing there may have been a connection between their personalities and the nature of their discoveries. Both of them understood that thinking is a mechanical process, and such understanding requires a change of position, from a participant in the game of thinking, to that of an onlooker. The detachment of the two may have had a part in their ability to change their position so. Somebody immersed in social interaction would find it hard to reach such ideas.

All this is said in the positive. Gödel's and Turing's way of looking at thought is the right one. To understand it well, without falling for the traps posed by circularity, it is necessary to look at it from afar.

Can a Machine Know that it is a Machine?

The classical Copernican revolutions have withstood the test of time; few believe nowadays that the sun circles the earth. Those who oppose Darwin's theory do it mainly out of religious motives, and the ideas of psychoanalysis have permeated modern thought so thoroughly, that even those who deny the existence of the unconscious unheedingly use Freudian concepts. Turing's Copernican revolution, by contrast, has not gone far. Turing succeeded in constructing thinking machines, but not in convincing his fellow men that man has no advantage over the computer. Most people still believe that the human mind is not part of the physical world.

Admittedly, this is not of major significance. To believe that the sun circles the earth or deny natural selection are both the result of ignorance, and ignorance has its price. Rejecting the Copernican revolution of Turing, on the other hand, has little practical implications. As we saw in the part about the Mind–Body problem, it results from lack of separation between the positions of observer and observed when viewing one's own mind. That is, it is the result of some knot that refuses to be untangled. Certainly not the end of the world.

And still, it is an intriguing question — why is it so hard for man to understand that he, including his thinking, is part of the physical world, no different from other parts? One answer to this question can be learned from Searle's story of the Chinese room. When a native Chinese speaker speaks Chinese, he has a powerful tool at his hands — a language that was assimilated over many years. When he relinquishes this position and looks at himself from aside, he is giving up this tool, and then he feels depleted. From aside, his behavior looks to him like empty gestures, those gestures that Searle ascribes to the raiser of boards with incomprehensible signs. This he may refuse to accept: is this me, he thinks, the one who is so good in communicating and playing the social game?

But possibly the deeper reason is that not only man cannot understand he is a machine. No machine can understand this, at least not when it acts. When you try to pin a person to a board like a butterfly, he is bound to protest: "But I can act also otherwise, not as you describe me." This is the protest of the described against the description. Machines made of steel and silicon will face the same limitation. Circularity presents man, like machine, with his limitations. And one of these limitations is the inability to be fully aware of them. Isn't this the ultimate victory of circularity?

Three Families, One Secret

> Three friends sit in a bar, and as often happens in such cases, a fairy appears, turns to one of them and offers him a choice between three wishes: beauty, a million dollars or wisdom. He thinks and thinks, and eventually chooses wisdom. The fairy waves her wand, and suddenly the man's friends see that he is crestfallen. "What's the matter?" they ask. "I should have chosen the million dollars."

Looking for a sure recipe for a funny story? Tell one in which an idea that seems to point at the outside world turns out to be pointing at itself, in other words a story of circularity. Circularity has a dark side and an illuminated one, but also a third side — a funny one.

> As long as Stalin was alive, every book published in Russia had to mention his name at least once. In one book the name Stalin appeared in the index, with reference to a certain page: this very page of the index.

The Monty Python group likes this type of humor. In *Life of Brian,* the group's version of the story of Jesus, Brian opens the window of his room in Jerusalem, and sees a multitude gathering in the street below, calling "A grace, a grace!" "You don't have to follow me," he calls, "you are all individuals!" "Yes, we are all individuals," they shout back in unison. "You are all different," he tells them, and they all call together, "We are all different!" A voice is heard from the crowd: "I am not!"

> A manager has a hundred CVs on his desk. He shuffles them well, and throws half to the trash bin. Asked why he did it, he answers: "I don't want losers."

And a less refined one:

> Interviewer: What is your worst shortcoming?
> Job applicant: Honesty.
> Interviewer: In my opinion, honesty is not necessarily a shortcoming.
> Job applicant: I don't care about your stupid opinion.

Mathematicians say about it a "Q.E.D." Proof done.

> He is such a total loser, that if a contest of losers were held he would come out last.

Why are these jokes funny? Aha, this is a very good question. The question "what is funny" is not simple at all. Generations of thinkers have been batting their heads against it. Circularity can help us here, since it is a well-defined mechanism, and is easy to recognize. But let us not attack the problem head on, and instead go circuitously. I want to discuss another family of jokes that has a sharp contoured mechanism: those based on detachment of an action from its intention.

X throws a pie at Y, Y bends and the pie hits Z. A man walks and slips over a banana peel. Why are these funny? "Surprise", goes a common explanation. But surprises are not always funny. "Derision", says an explanation of Plato and his disciple Aristotle, for whom "laughing" was synonymous with "laughing at", and therefore both advocated seriousness. But not all mishaps make us laugh. The common denominator to these two events is different: in both an action is detached from its intention. In the first the outcome is different from that which X intended, and in the second the person wanted to continue walking, and the banana peel intended otherwise. Detachment of intention is not less common in jokes than circularity.

> Two laborers toil on a mountain. One digs a hole in the ground, the other fills it, one digs, the other fills. An onlooker is puzzled, "What are you doing?" "Usually we are three," explains the digger. "I dig, Sasha plants a tree, and Misha refills with soil. Today Sasha is off sick."

The digging is detached from its purpose — the planting.

> An old lady returns to her room in the old age home to find another old lady with her hands over her husband's pants. She is furious, "What does she have that I don't?" "Parkinson's," answers the husband.

An action that is supposed to be loaded with meaning turns out to be involuntary convulsions, detached from drive. The next joke, too, detaches sex from drive:

> A man is stranded on a desert island and finds six women there. They make an arrangement: every weekday he does it with another woman, and Saturday is a day off. One day another man reaches the island. The first comer is pleased, "We can share the work." "Sorry, I am gay," announces the newcomer. "Shacks," says the guy, "there goes my day off."

Who wouldn't dream of six women on a desert island? But for the guy this is work. Here is a very similar one — sex as chore:

> "Now in, now out. Now in, now out," the farmer's daughter instructs the inexperienced farm boy. "Make up your mind," he tells her, "I must feed the cows."

Has the new family of jokes brought us nearer to the secret of jokes? Hardly. Nothing seems to be common to the two types. But in fact, they do share an underlying mechanism, for the discovery of which I will tell you about a third family of jokes. The name of this technique, "realization of a metaphor", is taken from the theory of poetry. It means taking the air out of the balloon of a symbol. Metaphors are taken at face value, and symbols turn out to be simple objects, devoid of meaning.

> Woman: I will never give you my heart.
> Suitor: I was not aiming that high.

The suitor takes the metaphor of "giving one's heart" literally, and then lowers it still further. This mode of thought is called "reification" ("*rei*" means "body" in Latin), or "concretization". Freud claimed that it is characteristic of schizophrenics, and nowadays it is ascribed to people with Asperger's syndrome.

> What is the epitome of wastefulness? — telling a hair-raising story to a bald man.

Sometimes the symbol that is emptied of meaning is familiar, and sometimes it is a symbol that the joke itself created.

> Two nonagenarians marry. On the first night he gropes for her hand, they hold hands and fall asleep. On the second night he gropes for her hand, they hold hands and fall asleep. On the third night he gropes for her hand, and she says, "No, darling, not tonight. I have a headache."

The holding of hands arouses anticipation of what must follow. On the third night it transpires that it is the thing itself.

In the next joke, too, a metaphor formed by the joke itself is flattened:

> The wife of the chief of a tribe in Africa gives birth to a white baby. The suspicion falls, of course, on the local missionary, and the chief takes him for a talk. "It is just a coincidence," says the missionary. "Here, look at the hill over there — there is a herd of white sheep, and one of them is black. A coincidence." "OK," says the chief. "You don't talk about my coincidence, I don't talk about your coincidence."

The sheep, that were meant only to be a metaphor, turn out to be more interesting than what they were supposed to symbolize.

> A Chinese couple makes a pact when they marry: each of them will have a jar, and whenever one is unfaithful to the other, the unfaithful partner will put one grain of rice in the jar. After fifty years of marriage they decide to open the jars. The husband's jar is found to contain three grains of rice.
> "What was the first?" asks the wife.
> "You remember when your mother was ill, and you went to nurse her for a few months? I did it with the young school teacher."
> "And the second?"
> "Do you remember the nice maid we had some thirty years ago?"
> "And the third?"

> "Do you recall the time of the big flood, when I went to the big city?"
> Then they open the wife's jar, and it is empty. "Have you never been unfaithful to me?" asks the husband.
> "Do you remember," says the wife, "the big famine, when everybody starved and we had plenty to eat?"

Rice, so it turns out, is not only a symbol, but also food.

The three families of jokes look very far apart. What does circularity have to do with detachment of intentions, or with the flattening of metaphors? In fact, they do have a common mechanism: a symbol is victorious over its meaning. In all three families a symbol sets itself free from the yoke of its meaning, and attains a life of its own. Here are two examples of such a victory:

> Could your honor the Prime Minister summarize the state of the nation? In one word — good. And in two words? Not good.

It is the number of words that matters, not the meaning. Form over content.

> A gregarious woman studies the restaurant menu and eventually says to the waiter, "Yes."

A menu is a symbol (yes, even menus can be symbols), in that it indicates the possible choices. Here it becomes the thing itself.

Most of their lives the symbols are servants of meaning. "Good prose is like a window pane," said George Orwell, and meant that the words should be transparent, the listener being aware only of their meaning. When we use symbols, we rarely think about their external shell. But sometimes the wheel turns. This happens, for example, in poetry. Jean Paul Sartre added to Orwell's dictum (possibly without even knowing it) "Prose is like a window pane; poetry is like a mirror." In poetry the words are no longer transparent. We are here, they rebel, not merely servants of what we mean. We have an existence of our own. And this happens also in jokes.

Let us examine this mechanism in the above three families of jokes. The family in which it is most conspicuous is the realization of metaphors. When taken at face value, a symbol remains an empty vessel, and our attention switches from the meaning to it. "Look at the jar not at what's inside it".

More enigmatic are the jokes in which actions are detached from their meaning. Where is the "symbol" in these jokes? The secret is that "symbol" should be understood in a wide sense, as everything that directs our attention to another thing, and then this other thing is the meaning. In this sense, an action is a symbol, since we attribute meaning to it. We ask people for "the meaning of their actions", where the meaning can be aim, or purpose, or cause, or drive. Understanding the goals and intentions behind our fellow men's actions is vital to our survival, which is the reason why we are always busy deciphering them. In jokes of detachment of intentions the action remains an empty shell — the symbol, in this case the action, vanquishes its meaning.

So, finally we reach our original goal: what makes circularity funny? Circularity is a classic case of the victory of the symbol. A symbol that seems to point at the world turns out to point at itself, thereby getting the upper hand over the meaning. Can there be a clearer case of the victory of the symbol?

Circular humor often plays on a theme that has recurred throughout our discussion: the difference between being inside the game as opposed to looking at it from the outside. For example, this is the trick in "trap jokes", in which the listener ends up as part of the joke.

> Want to hear a joke from end to beginning? — Yes! — Then laugh first.

> Do you want to hear a joke? — Yes. — OK. Once there were two jokes. One fell ill. The other sat by its bed, supported it, fed it soup. Isn't it a good joke?

Though it is a children's joke, it is quite sophisticated. There is a quick transition from being outside to being inside the joke — you realize that "a good joke" relates to the one you are presently hearing, and you have to admit that indeed, it is a good joke.

There are jokes in which the protagonist discovers that he is the subject of the words whose meaning he is pondering:

> A man walks along the wall of a mental asylum, and from the inside he hears, "Seven! Seven! Seven!" His curiosity aroused, he

> climbs and looks over the fence. He receives a blow to his head, and hears the inmates shout, "Eight! Eight! Eight!"

The preference of the symbol over its meaning is expressed in a preference of thought over its content. A famous example is the cartoon figures, who continue to walk after the ground has disappeared beneath their feet, and fall only when they realize there is an abyss below. The recognition of the abyss is what matters, not its existence. Here is the same idea in a saying of Mark Twain:

> When I was a boy of fourteen, my dad was so ignorant I could hardly stand to have the old man around. But when I got to be twenty one, I was astonished at how much he had learned in seven years.

And here it is in a joke:

> Wife: You must stop drinking. We are out of money.
> Husband: Just yesterday you spent $200 on make-up!
> Wife: This is so that you would find me pretty.
> Husband: That's what the beer is for.

Finally, a natural question should be answered: do all jokes contain a grain of circularity? Probably not. But circularity is more prevalent in jokes than meets the eye. Look for example at the following:

> A tourist is watching two cows in the meadow, one white and one black. A farmer stands next to him, and the tourist asks politely, "These cows, do they produce a lot of milk?"
> "The white cow does," answers the farmer.
> "And the black?"
> "The black one too."
> After some silence the tourist asks, "Do they give birth to calves every year?"
> "The white cow does," says the farmer.
> "And the black?"
> "The black does too."
> This kind of exchange goes on for a while, until finally the tourist gets curious, "Why do you always answer on the white cow first and only then on the black?"
> "You see," says the farmer. "The white cow is mine."
> "And the black?"
> "The black one too."

The joke returns to its starting point, as an automaton that returns the coin inserted into it. Lots of jokes are constructed like a boomerang that is flung into the world only to fly back at its sender. And, speaking of boomerangs:

> Did you hear? Joe is in hospital.
> What happened to him?
> He got a new boomerang and threw out the old one.

The Jail of Personality

Now that I have finished speaking, I would like to add a few words.

(Conclusion of a speech)

We have met a few facets of circularity: it is a villain that wastes people's time and diverts them from the straight and narrow; a source of understanding the boundaries of knowledge; a mathematical technique that is responsible for deep results; a sure recipe for humoristic effect. Have we missed anything?

Indeed we have, and a central aspect at that: the role of circularity in everyday life. We are not always aware of it, but it is always there, in every living moment, on any occasion that we think about ourselves and about our goals in life. A concise way of saying this is: when we think about our motives, the thought itself emanates from the very same motives. And usually the thought is more interesting than its content matter.

For example, if a person rebukes himself for trying to ingratiate, be assured that behind the scolding stands the very wish to ingratiate. "If only I will not try to be liked, I will be liked". Sounds like a circularity based joke? Indeed, but this is also part of life. When a person reproaches himself, he constructs in himself an imaginary parent figure, that in some fantasy promises him love if he just behaves properly. Once he has done so, he feels he is now in a different position, blameless and strong and immune to criticism. But this is just fantasy. You cannot get away from your personality.

"The prisoner cannot free himself from jail", goes a saying in the Jewish *Talmud*. This motive has preoccupied philosophers of all times.

Plato spoke of the jail of knowledge, simply meaning that you cannot know what you don't know. In his famous cave parable he compared humans to people sitting in a dark cave, watching on the inner wall the silhouettes of figures that walk outside. The cave inhabitants will never experience anything but the shadows, and will never be able to know what the real figures are. Descarte's demon, whom we met in the chapter The Cat Sits, Therefore the Cat Is, is another instance of a jail story — that of the sensations. You will never know what is really beyond them. Escaping the jail of consciousness is a circular task, which in contrast to escaping physical jails (that is sometimes possible, as the Count of Monte Cristo can testify) is never possible.

The fear that we shall never know what is really out there in the world, because we cannot get out of our skin, we called "circularity anxiety". Wittgenstein expanded this anxiety to language. Since our consciousness is molded by language, and since we only think using the words we know, we are words' captives. In *Philosophical Investigations* he writes:

> A picture held us captive. And we could not get outside it, for it lay in our language and language seemed to repeat it to us inexorably.

But in everyday life all these are dwarfed by a much more significant jail: that of our personalities. Nobody can change his spots, outrun himself or go out of his skin. Our personality crystallizes over such a long period, and has such deep layers, that we cannot change it in one fell swoop. The one hundred billion synapses in our brain will not easily change their links by one decision.

The strange thing is that we cannot see this. We believe that we can create ourselves anew at any moment. "You can make a fresh start with your final breath," wrote Bertolt Brecht in his poem *Everything Changes.* And as is often the case with poems, there is inner truth in this, literally "inner": looking from the inside, this is true. He who acts and decides cannot fully understand the laws governing his behavior. From the inside, we believe that we can change everything at any given moment. A machine cannot understand it is a machine. And so, while man cannot get away from the jail of his personality, neither can he be fully aware of the existence of the jail, not as long as he deliberates and decides. And it is good that this is so. It is a good thing that we think that we can change everything until our last breath.

All along the book we have seen that the dark and illuminated sides of circularity meet and intermingle. Confusion that it sows in one place generates insights in other places. The same is true here. On the one hand, when we are in the midst of action, a restriction on self-knowledge prevents us from fully knowing our deep motives. On the other hand, understanding that our personality has depth, and that each of us is a prisoner of his personality, is an essential condition for self-awareness. And if this sounds paradoxical, it is because circularity is here at play.

Part VII: For the Experienced Hikers

The Diagonal Method

Tell all the truth but tell it slant —
Success in Circuit lies

(Emily Dickinson)

Cantor arrived at his theorem in two stages. In the first stage, he proved only that not all infinite sets are of equal size. As already mentioned, the smallest infinite set is the set of natural numbers, and Cantor proved that there is a larger set. In other words, **there are sets that cannot be counted**. "Uncountable" they are called.

Cantor's proof is explicit. Not only does it show that there is an uncountable set, it also expressly presents such a set. This is the set of sequences composed of 0s and 1s. There are many such sequences. For example: 0, 0, 0, 0, ...; 1, 1, 1, 1, ...; 0, 1, 0, 1, 0, 1, ...; or 0, 1, 0, 0, 1, 0, 0, 0, 1, 0, 0, 0, 0, 1, ... Cantor showed that there are not just many sequences like these, there are very many, more than natural numbers.

The proof is by negation. That is, upon assuming that the theorem is false, a contradiction is reached. Let's assume, Cantor said, that it is possible to count all the sequences whose terms are 0 and 1. The enumeration below is just an attempt, whose purpose is to make things concrete:

$S_1 = 0, 0, 0, 0, 0, 0, 0, \ldots$
$S_2 = 1, 1, 1, 1, 1, 1, 1, \ldots$
$S_3 = 0, 1, 0, 1, 0, 1, 0, \ldots$
$S_4 = 0, 0, 1, 1, 0, 0, 1, \ldots$

$S_5 = 1, 0, 1, 1, 0, 1, 1, \ldots$
$S_6 = 1, 0, 1, 0, 1, 0, 1, \ldots$
$S_7 = 0, 0, 0, 1, 1, 1, 0, \ldots$

The assumption is that all 0, 1 sequences should appear in this list. Namely, that we succeeded in counting all of them. But now, Cantor said, I will show you that in actuality you must have failed to count them all. I will show you a sequence that definitely does not appear in this list. To this end, look at the diagonal of the table:

$S_1 = \mathbf{0}, 0, 0, 0, 0, 0, 0, \ldots$
$S_2 = 1, \mathbf{1}, 1, 1, 1, 1, 1, \ldots$
$S_3 = 0, 1, \mathbf{0}, 1, 0, 1, 0, \ldots$
$S_4 = 0, 0, 1, \mathbf{1}, 0, 0, 1, \ldots$
$S_5 = 1, 0, 1, 1, \mathbf{0}, 1, 1, \ldots$
$S_6 = 1, 0, 1, 0, 1, \mathbf{0}, 1, \ldots$
$S_7 = 0, 0, 0, 1, 1, 1, \mathbf{0}, \ldots$

Now, write the sequence that appears on the diagonal: $S = 0, 1, 0, 1, 0, 0, 0, \ldots$ and change every 0 in it to 1, and every 1 to 0. This gives us the sequence $T = 1, 0, 1, 0, 1, 1, 1, \ldots$. What is special about the sequence T? It differs from S in each of its terms. Since the first term in S is the first term in S_1, we know that T differs from S_1 in its first term: S_1 has 0 in the first place, while T is defined as 1. Consequently, T is not equal to the sequence S_1. Similarly, T is different from S_2 in its second place: S_2 has 1 in its second place, and T reversed it. It has 0 in the second place. And if T is different from S_2 in the second place, the two cannot be the same sequence (if two sequences are equal, they are equal in all places). T differs from S_3 in its third place, and therefore is not identical to S_3. And on and on: for each number i, the sequence T is different from the sequence S_i in the ith place. Accordingly, T differs from all the sequences S_i, that is, T does not appear in the list. So, this list does not include all counted 0, 1 sequences in the world! And this is true for every attempt to count the 0, 1 sequences. Every such attempt is doomed to failure.

Conclusion: The Real Numbers are Not Countable

One of the conclusions of this theorem is that there are more real numbers than natural numbers. Every sequence of 0s and 1s can be matched with a real number, by adding a 0 and a decimal point to the left of the sequence. For example, the sequence 0, 1, 0, 1, ... corresponds to the number 0.0101...; the sequence 0, 0, 1, 1, 0, 0, 1, 1, ... corresponds to the real number 0.00110011... . Since the sequences whose terms are 0 and 1 cannot be counted, it is also impossible to count all the real numbers of this type, that is, the numbers in whose decimal representation a 0 appears before the decimal point, and only 0s and 1s after it. These, of course, are only a small portion of all the real numbers, and if even these cannot be counted, then all the real numbers most certainly are uncountable.

For obvious reasons, the method used to prove this theorem is called the "diagonal method". Since its discovery, it has repeatedly proved its effectiveness, and has become a standard mathematical tool.

A Methodical but Not Very Efficient Detective

As recalled, one of Hilbert's challenges to the mathematical world was finding an algorithm that tests for provability. Gödel's Incompleteness Theorem, so we realized, entails that this task is impossible. Let us show now the converse: that if you know that this challenge is impossible, then the Peano Axiom system is incomplete.

The secret is in a natural candidate for an algorithm that tests provability. It consists simply of trying all the possibilities for a proof. Given a formula you want to check for provability, go systematically, from shorter to longer, over all possible sequences of symbols. For each such sequence, check to see whether it proves the formula or not. Most of the sequences are not proofs at all, but just heaps of disconnected symbols. But perhaps, by chance, like the monkey hitting a typewriter at random, you will hit upon a proof of the formula. If there is such a proof, you will get to it at some point. And since checking whether a sequence of symbols is a proof of a given formula is doable, the algorithm is well defined.

Obviously, this isn't very efficient. It's like a police detective trying to solve a murder case by examining all the people in the world, one by one. Just as police investigations aren't conducted in this manner, so, too, no one would try to find proofs for mathematical theorems by writing random marks on paper and then checking whether, by chance, they provide a proof for the theorem. But at this stage we aren't looking for an efficient algorithm, but for any algorithm at all.

But there is a worse and deeper problem. It is that we don't know when to stop. The algorithm of the murder detective isn't efficient, but it is possible, since there are only a finite number of people in the world, and at some point the algorithm will end. The situation is different for mathematical proofs. If a proof is found at some point during our search, well and good, but what will we do if we have examined a million series of signs, and none of them is the proof we seek? Obviously, we could continue on to the one-million-and-first series, but we would never be able to stop and declare: "We exhausted all possibilities, and have not found a proof, hence the statement is incorrect." There is always the possibility that in our next step we would discover the proof, if only we were to continue. Oil prospectors face this dilemma, but in their case there is at least a theoretical limit: if they drilled and reached the other side of the earth, this is a clear sign of failure. As far as proofs are concerned, there is no phase in which we should give up. There is no theoretical bound to proof length.

But now assume that PA is complete. Then, either our formula is provable, or its negation is. So, change a bit your algorithm: for each sequence of symbols you have jotted, check if it is a proof of the formula, and if it is not then *check if it is a proof of the negation of the formula.* If PA is complete, then one of the two possibilities will occur at some point. This means that with certainty the procedure will end at some point. So, if PA is complete, we have at our hands an algorithm for checking provability. Knowing that there is no such algorithm entails therefore that PA is not complete.

Why Gödel's Paradox Doesn't Generate a Mathematical Contradiction

Like all other paradoxes, Gödel's paradox is based on some deception. To realize what it is, it is best to scrutinize its mirror image, Gödel's Incompleteness Theorem, and see why it doesn't lead to a contradiction. We suspected that the cheat is in the use of the "axiom" saying that everything provable is correct. This is the link in the chain of arguments that is missing in the mathematical proof. There just isn't any such rule in logic.

But we could try to add this rule. Why not? After all, it should be true, if our proofs are based on sound principles. How can we do it? We should only recall the formula $\pi(x)$, that says about the number x that it is the number of a provable formula. So, we could add, for every formula α, the axiom $\pi(n) \rightarrow \alpha$, where n is the number of α. This means: "If α is provable then it is true" — a very natural axiom, isn't it? This will demand adding infinitely many axioms, but this is fine. Systems of axioms with infinitely many axioms are permitted. In fact, every Peano system contains infinitely many axioms, because one of the axioms, about the principle of induction, is in fact a pattern that generates infinitely many axioms. Having added these new axioms, we could repeat Gödel's proof of the Incompleteness Theorem, and then adding this axiom at the final stage would produce a contradiction, just as it does in the paradox!

Fortunately, this doesn't work. This argument conceals a deceit, and as could be expected — a circular one. The formula $\pi(x)$ is dependent on the system of axioms. For every system of axioms there is a different formula expressing provability in the system. So, adding new axioms should change the formula $\pi(x)$. But this hurls us into a vicious circle: having changed the formula we should add new axioms of the form: $\pi(n) \rightarrow \alpha$, which will force us to change again the way $\pi(x)$ is constructed, and so on, *ad infinitum.*

The failure to formulate a mathematical version of the claim "if something is provable then it is true" is the best testimony to the location of the flaw in Gödel's paradox. We have seen that any attempt to add such an axiom is doomed to failure, being a circular task.

Why the Incompleteness Theorem Doesn't Contradict the Completeness Theorem

Russell and Whitehead's 1913 *Principia* was a great gift to mathematics. It presented a simple and well-defined system of proof. But was it the last word? Namely, can we be certain that it can prove every true fact? That it does not miss any rule of inference?

This question was open for almost twenty years. In hindsight, the difficulty was not in its complexity, but in the young age of the concepts of mathematical logic. What is a "true fact"? And what does it mean that "every true fact can be proved"? A person with a strong torch was needed to clarify this. Indeed, such a person came — Kurt Gödel, and his torch was a scorchingly bright mind. A great advantage in mathematics, less so in life. Few people ever roamed earth that were more rationalistic. This caused him pain in his personal life, but mathematics benefited. He proved that indeed the Frege–Russell–Whitehead system is powerful enough to prove every true fact. This is the Completeness Theorem. It is of major importance for mathematicians: they can now be sure that every question has an answer. Every conjecture can be proved or refuted. If a fact F cannot be proved from a system of axioms S, then there exists a structure satisfying the axioms of S but not F.

Confusingly, a year later Gödel proved the Incompleteness Theorem. But this does not contradict the Completeness Theorem, since the incompleteness is of something else — the Peano system of axioms. He found a formula G that is true in the natural numbers, but cannot be proved or from these axioms. This does not contradict the Completeness Theorem, since by the latter the fact that G cannot be proved from the axioms only says this: that there exists a structure in which Peano Axioms are valid, and G is not. Indeed, G is true — but only in the natural numbers, not in any structure. Peano Axioms do not determine everything, they do not force a structure satisfying them to be precisely the natural numbers. There exists a structure satisfying them that is not the natural numbers, and G, though true in the natural numbers, is false in that structure.

Bibliography

Part I: Magic

Barwise, J., and Etchemendy, J. (1987). *The Liar.* New York: Oxford University Press.

Hughes, G. E. (ed.). (1992). *John Buridan on Self-Reference.* Cambridge and New York: Cambridge Univesity Press.

Kripke, S. (1975). "An outline of a theory of truth", *Journal of Philosophy,* 72 (19): 690–716.

Lefebvre, N., and Schelein, M. (2005). "The liar lied", *Philosophy Now,* 51.

Priest, G. (1984). "The logic of paradox revisited", *Journal of Philosophical Logic,* 13: 153–179.

Smullyan, R. (2011). *What is the Name of this Book?* Mineola, NY: Dover Recreational Math.

Part II: Free Will

Callender, C. (2001). "Thermodynamic asymmetry in time", in: E. N. Zalta (ed.). *The Stanford Encyclopedia of Philosophy.*

Campbell, R., and Sowden, L. (eds.). (1985). *Paradoxes of Rationality and Cooperation: Prisoner's Dilemma and Newcomb's Problem.* Vancouver: University of British Columbia Press.

Danto, A. C. (1968). *What Philosophy Is.* New York: Harper and Row.

Darrow, C. (1989). "The plea of Clarence Darrow, in defense of Richard Loeb and Nathan Leopold, Jr., on trial for murder", in: S. Cahn (ed.), *Philosophical Explorations: Freedom, God, and Goodness.* New York: Prometheus Books.

Gardner, M. (1974). "Mathematical games", *Scientific American,* 230(3), p. 102.

Gardner, M. (1986). *Knotted Doughnuts and Other Mathematical Entertainment.* New York: W. H. Freeman, pp. 15–17.

Levi, I. (1982). "A note on Newcombmania", *Journal of Philosophy,* 79: 337–342.

Nozick, R. (1969). "Newcomb's problem and two principles of choice", in: N. Rescher (ed.), *Essays in Honor of Carl G. Hempel.* Dordrecht, The Netherlands: D. Reidel, pp. 114–146.

Penrose, R. (1999). *The Emperor's New Mind.* Oxford, New York, Melbourne: Oxford University Press.

Schlick, M. (1984). *Fragen der Ethik.* Frankfurt am Main: Suhrkamp.

Strawson, G. (1998). "Free will", in E. Craig (ed.), *Routledge Encyclopedia of Philosophy.* London: Routledge, Vol. 3, pp. 743–753.

Part III: The Mind–Body Problem

Cavell, S. (1969). *Must We Mean What We Say: A Book of Essays.* Cambridge and New York: Cambridge University Press.

Euler L. (1802). *Letters of Euler on Different Subjects in Physics and Philosophy Addressed to a German Princess.* Translated from French by Henry Hunter, D. D. London: Murray and Highley, Letter 3.

Popper, K. R., and Eccles, J. C. (1977). *The Self and its Brain.* Berlin/Heidelberg/London/New York: Springer-Verlag.

Ryle, G. (1950). *The Concept of the Mind.* Chicago: University of Chicago Press.

Part IV: Large Infinities and Still Larger Ones

Clegg, B. (2003). *Infinity – The Quest to Think the Unthinkable.* London: Robinson.

Part V: Gödel's Incompleteness Theorem

Goldstein, R. (2006). *Incompleteness: The Proof and the Paradox of Kurt Gödel.* New York: W. W. Norton.

Hofstadter, D. R. (1980). *Gödel, Escher, Bach: an Eternal Golden Braid.* New York: Vintage Books.

Smullyan, R. (1994). *Diagonalization and Self-Reference.* Oxford: Oxford University Press.

Sokal, A., and Bricmont, J. (1998). *Intellectual Impostures.* London: Profile Books.

Part VI: Turing Invents the Computer

Hodges, A. (2014). *Alan Turing: The Enigma.* London: Vintage, Random House.

Hofstadter, D., and Dennet, D. (1982). *The Mind's I: Fantasies and Reflections on Self and Soul.* New York: Bantam Books.

Searle, J. (1980). "Minds, brains and programs", *Behavioral and Brain Sciences,* 3: 417–457.

Searle, J. (1999). "The Chinese room", in: R. A. Wilson and F. Keil (eds.). *The MIT Encyclopedia of the Cognitive Sciences.* Cambridge, MA: MIT Press.

Freud, S. (2003). *The Joke and its Relation to the Unconscious.* New York: Penguin.

Printed in the United States
By Bookmasters